THE BOOK OF
FRUIT AND FRUIT COOKERY

PAUL DINNAGE

The Book of

Fruit and Fruit Cookery

Illustrations by Meg Rutherford

Sidgwick and Jackson
London

First published in Great Britain in 1981
by Sidgwick and Jackson Limited

ISBN 0–283–98553–4

Typeset by Computacomp (UK) Ltd, Fort William, Scotland
Printed in Great Britain by
The Garden City Press Limited
Pixmore Avenue, Letchworth, Hertfordshire SG6 1JS
for Sidgwick and Jackson Limited
1 Tavistock Chambers, Bloomsbury Way
London WC1A 2SG

ACKNOWLEDGEMENTS

This work could not have been completed without extensive reference to earlier cookery books and books about food. Of general help have been *A Concise Encyclopaedia of Gastronomy* (André Simon, 1942), the original French edition of *Larousse Gastronomique*, Theodora Fitzgibbon's *The Food of the Western World* (1976), Miss Dorothy Hartley's *Food in England*, Dr E. Lewis Sturtevant's *Notes on Edible Plants* (1919), Popenoe's *Manual of Tropical and Sub-tropical Fruits* (1923), Mrs M. Grieve's *A Modern Herbal* (1931) and similar works.

I am also grateful for ideas derived from the published works of a number of authors who collected recipes in the period between the wars: they include A. H. Adair, X. M. Boulestin, and Ambrose Heath, and (among the many ladies who followed the example of Eliza Acton) Dorothy Alhusen, May Byron, Lady Clark of Tillypronie, Oriana Haynes, Mrs C. F. Leyel, Ruth Lowinsky, Mrs Phillip Martineau, and the estimable Miss Florence White. Among contemporary authors, I am indebted to Mrs Robin Howe, Miss Gertrude Mann and the late Rupert Croft-Cooke.

Paul Dinnage

CONTENTS

WEIGHTS, MEASURES, TEMPERATURES

The measurements in this book are British: teaspoon and tablespoon measurements are flat unless specifically described as heaped.

Oven Temperatures

Description	*Deg. F*	*Deg. C*	*Gas Mark*
Very Cool	225	110	¼
	250	120	½
Cool	275	140	1
	300	150	2
Moderate	325	160	3
	350	180	4
Moderately Hot	375	190	5
	400	200	6
Hot	425	220	7
	450	230	8
Very Hot	475	240	9

It should be noted that the real heat in a gas or electric oven varies according to size and its own characteristics.

Quantities

Ounces,	*pounds*	*Metric equivalent grammes*	*Suggested metric equivalent*
1	–	28	25
2	–	57	50
3	–	85	75
4	–	113	100
5	–	142	150
6	–	170	175
7	–	198	200
8	½ lb	227	225
12	¾ lb	340	350
16	1 lb	454	450
	2 lb	908	1 kg

Weights, Measures, Temperatures

Liquid

Pints	*Metric equivalent millilitres*	*Suggested metric equivalent*
1/4	142	150
1/2	283	300
3/4	425	450
1	567	600
1 1/2	851	900
1 3/4	992	1 litre
2	1134	1·2 litres

Spoon measurements

Teaspoon

AMERICAN	BRITISH	AUSTRALIAN
1 = 5 ml	1 = 5 ml	1 = 5 ml
Tablespoon		
1 = 14·2 ml	1 = 17·7 ml	1 = 20 ml

One heaped American tablespoon is held to be the same as one level British tablespoon.

Fluid measurements

American pint	=	16 fluid oz
British pint	=	20 fluid oz
American cup	=	8 fluid oz
British cup	=	10 fluid oz

PASTRY-MAKING

Fruit tarts, pies, puddings and other fruit pastry dishes call for a basic knowledge of pastry-making. This relies on simple ingredients, comparatively simple methods with certain rules, and, it is to be hoped, some of the flair that produces good pastry. The French do not tire of reminding those who cook that a good *pâtissier* is born and not made; however, practice can produce excellent results.

INGREDIENTS

Flour	should be fine, white flour, perfectly dry, and sieved. Self-raising flour is not advised, except for suet pastry.
Shortening	or fat should be either butter (for puff pastry and shortcrust pastry), lard (for shortcrust, sometimes mixed with butter), beef dripping, or clarified beef suet. Margarine, bacon fat, mutton dripping are not recommended, although pastry of a kind could be made with these fats.
Water	the only liquid used in making most pastry, should be cold.
Salt	is added to all pastries, a generous pinch to every ½ lb (225 g) flour.
Sugar	must be castor sugar, and preferably used sparingly.
Lemon juice	squeezed from half a lemon is a counter-agent to richness, and improves the flavour slightly.
Egg yolk	is added for crispness and shaping quality, and to give a richer colour, 1 egg yolk to each 6 oz (175 g) flour.

RULES FOR PASTRY-MAKING

Have everything quite clean, cool, and completely dry – this applies to the pastry board (it must be wood, or ideally the old-fashioned marble slab), rolling pin, wooden spoons, flour dredger, mixing bowl, pie dishes and baking trays.

Use the coldest available water, and cold, but not over-chilled, butter. Sift the flour to get air into it.

Make pastry in the coolest place possible, and, as pastry is a hand operation, keep your hands cool.

Avoid indifferent pastry by using as little liquid as possible and by sticking to fixed baking times: adding too much water and over-cooking are two common faults.

RICH SHORTCRUST PASTRY *(pâte brisée)*

For pies and tarts, an all-purpose shortcrust pastry can be made as follows, the proportions being constant whatever the quantity.

Rub 4 oz (100 g) butter into 6 oz (175 g) plain flour, and add $\frac{1}{2}$ teaspoon salt. Mix well together. With the addition of a little cold water – not much more than a tablespoon – quickly make a firm dough. A little lemon juice may be squeezed in. Mix with the yolk of an egg and a little water, if a richer pastry is desired. An alternative fat mixture is 2 oz (50 g) butter and 2 oz (50 g) lard. Omitting the lard and adding an egg makes a pastry of firmer texture. Make sweet pastry, *pâte brisée sucrée*, by adding 1 oz (25 g) castor sugar. Pastry cases should be rolled out on a lightly and evenly floured board. Roll out to an even thickness. First, however, prepared pastry must have time to rest somewhere cool, e.g. in a plastic bag in the refrigerator, for half an hour or more. This removes the elastic quality of the pastry, which would otherwise shrink back when baked. Cut the pastry to size with a sharp floured knife before baking. Grease the tray or tin for shortcrust pastry, plain or sweet.

Detailed instructions for the following pastries may be found in any comprehensive cookery book.

SUET PASTRY

When the fat used is suet, pastry needs either self-raising flour, or a mixture of baking powder and plain flour, to do the work that air does in other pastries. Finely grated fresh beef suet is used in the proportion of one third to one half of fat to one of flour.

PUFF PASTRY *(pâte feuilletée)*

This differs from shortcrust in being composed of many thin layers of pastry between which air is trapped. Only butter, preferably unsalted, is used to make puff pastry, and in a quantity equal to that of the flour, which should be the best available.

FLAKY PASTRY

Combines the shortcrust and puff pastry methods of making pastry. It is less rich than the latter, and particularly good for sausage rolls, meat pies, etc.

BAKING BLIND *(à blanc)*

Flans and tarts often need to have the pastry shell cooked for longer than its eventual content; the shell is therefore cooked 'blind', or empty, before receiving its fruit filling and being returned to the oven for the last stage of cooking. The pastry to be used is rolled thinly – about ⅛ inch (¼ cm) – and with an inch (2½ cm) to spare to allow for the sides, and pressed into position. The pastry should be pricked with a fork to prevent the bottom rising. Collapse of the sides of the pastry is prevented by lining with greaseproof paper and filling with dried haricot beans, uncooked rice, or a similar shaping agent that can be removed after the cooking time: for a 10-inch (25-cm) flan shell this would normally be 10 minutes in a moderately hot oven (400°F/200°C/mark 6), by which time the pastry will begin to take on its golden-brown colour. If the pastry is not going to be cooked any more when the filling is added, remove the paper and beans or rice and cook for a further 10 minutes.

Almond

THERE is no nut so valuable or so versatile as the almond. The protein value is higher than in any other nut, it contains iron and calcium, and it is an alkaline-forming food; the almond yields an excellent oil, and the emulsion – almond milk or cream – is a good digestive. The Romans acquired the tree from the Greeks, and may have introduced it to Britain, where it remained uncultivated until the mid-sixteenth century, and was then grown for the blossom rather than the nuts.

It is indeed a notably attractive tree, in growth and blossom not unlike the peach, to which it is related. The main difference is in the fruit, green and leathery in the almond, little more than a covering for the stone, luscious in the peach. But the nuts have been eaten, raw or sugared, by the Romans and others for centuries, and the lavish use of almonds in mediaeval cookery made the nuts an important article of commerce for many years. The tree, *Prunus dulcis (P. amygdalus* Stokes *var. dulcis* and *var. amara)*, conveniently grows all round the Mediterranean, all over Asia Minor and in China, and the almond nut is implicit in Middle Eastern, Greek, Italian, Spanish and Chinese cookery.

Apart from simply cooking almonds with vegetables and white meats – chicken and veal go well with almonds, and fish such as trout and sole, and crab and lobster, accept roast or blanched nuts – almonds have been used in great quantities by confectioners and pastry-cooks as flavouring. From mediaeval times onwards much effort went into the making of marchpane (later known as marzipan, the German term being adopted in the nineteenth century when the bulk of it was imported from Germany). Cakes of marchpane were invariably among the desserts at Elizabethan meals and feasts, and marchpane, sometimes not simply almonds and sugar, but with pistachio nuts, orange-flower water, white of egg and other ingredients mixed in, was used in 'comfits' and for ornamental sweets, to be coloured or even gilded; it was used on occasions for fantastic sculpture, later to be displaced by spun sugar. Almonds also form the basis of French nougat and the Spanish *turron.*

There are many varieties of almond: first, the sweet almond. (1) Jordan: the best (from Malaga – the name is a corruption of the French *jardin*); they are large, narrow, thin-skinned and quite unlike other almonds, and are to be used in all recipes unless otherwise specified; (2) Valentia: broad, with a thick dusty skin; (3) Sicily and (4) Barbary: both similar to Valentia but smaller. Secondly, the bitter almond. There are several kinds, from Sicily and the South of France; they are useful in small quantities for flavouring, but toxic if eaten in large quantities, as the nuts contain amagdalin, from which prussic acid may be extracted. Use bitter almonds only when they are specifically mentioned.

SALTED ALMONDS

Blanch ½ lb (225 g) whole almonds by pouring boiling water over them, leaving them a few minutes, then peeling them. Dry them in a cloth, and cook in a frying pan in which 2 oz (50 g) butter has been melted. Stir with a wooden spoon over moderate heat until a golden brown. Drain on paper. Sprinkle with table salt before serving in glass dishes.

DEVILLED SALTED ALMONDS

Blanch ½ lb (225 g) Jordan almonds (see recipe for salted almonds), dry them, then cook them in 2 oz (50 g) butter until golden brown, stirring slowly as this is done. Mix ½ tablespoon table salt with 1 tablespoon celery salt and a generous pinch of cayenne pepper. After draining the almonds, toss them in this mixture and serve in glasses.

The Polish have an almond soup, *zupa migdaowa*:

ALMOND SOUP

Mix ½ lb (225 g) blanched sweet almonds and 12 blanched bitter almonds. Have 3 pints (1·8 litres) milk ready, and use some of it in grinding the almonds to a paste. Pass this through a sieve with a little hot milk. The remaining milk is then brought nearly to the boil and mixed with the almond paste. Stir the mixture thoroughly and add sugar to taste. Heat until the soup bubbles up once, add 3 oz (75 g) rice, which has been cooked separately, and 2 tablespoons seedless raisins and serve.

This version of almond soup comes from Majorca:

SOPA DE ALMENDRAS Á MALLORQUIÑA

Chop ½ lb (225 g) blanched almonds; reserve 2 oz (50 g), chop them and pound the rest in a mortar, with 2 tablespoons cold water, into a fine paste. Take 2 pints (1·2 litres) chicken broth, add 2 tablespoons finely chopped lean ham, a pinch of ground clove, ¼ teaspoon mace and the same amount of nutmeg, and salt and cayenne pepper to taste. Add a bayleaf, and a pinch of thyme and basil. Simmer all these ingredients slowly for 20 minutes, then strain through a fine sieve. Add the pounded almonds, simmer a further 20 minutes, then stir in the chopped nuts. Boil up once. Stir in 1 tablespoon sherry and serve very hot with a spoon of whipped cream on each serving.

ALMOND PASTE BALLS FOR FRUIT SOUPS

Take 12 blanched sweet almonds and pound them into a paste. Beat 1 egg, and add it to the almonds with 3 oz (75 g) flour and 1 teaspoon baking powder. Mix, then roll the paste into balls the size of small plums. If the soup they accompany is hot, drop them in 5 minutes before serving; if it is a cold soup, cook the balls for 20 minutes in a very slow oven (275°F/135°C/mark 1) and serve separately.

TRUITE AUX AMANDES

Clean fresh brook trout, about ½ lb (225 g) each in weight, make an incision down the backs, and lightly rub with pepper and salt. Fry the trout in butter for 4 minutes each side in a heavy frying pan.

When ready, arrange on a hot plate. Return the pan to the heat, and as the butter becomes hot throw in slivered blanched almonds, sufficient to cover the fish completely. Agitate the pan to avoid burning, and when the almonds are a golden brown add 4 tablespoons dry white wine. Cover the fish with the almonds, add a few slices of lemon, and serve immediately.

OEUFS AUX AMANDES

Boil ½ lb (225 g) rice in plenty of water with ½ teaspoon saffron powder, a stick of cinnamon and a pinch of salt. When cooked, run hot water through it in a colander to separate the grains. Meanwhile, fry 20 blanched almonds and 1 oz (25 g) ground almonds in butter until they are a golden brown. Mix the almonds and the rice together, and make a pile of it on a dish. Poach 6 eggs, and place these round the rim of the dish. Cover them with a Bechamel sauce which has been enhanced with a little saffron and with English mustard. All the components of the dish should be equally hot when it is served.

This traditional regional recipe dates from 1741:

IPSWICH ALMOND PUDDING

Take 1½ oz (40 g) breadcrumbs and pour over them ¾ pint (450 ml) cream that has been warmed; into this stir 2 oz (50 g) castor sugar, 4 oz (100 g) ground almonds, and a little rose water or orange-flower water as flavouring. Beat up the yolks of 4 eggs and the whites of 2 eggs and mix well with the other ingredients. Pour into a buttered pie dish, putting a few pieces of butter on the top. Bake in a slow oven (300°F/150°C/mark 2) for about 45 minutes, and serve hot.

ALMOND MENCHIKOFF

Pound 4 oz (100 g) blanched almonds, and mix them with 4 oz (100 g) castor sugar and 4 oz (100 g) butter until they form a cream. Then add ¼ pint (150 ml) double cream that has been flavoured with vanilla. A smooth and consistent texture is essential. Turn into a mould that has been lined with boudoir biscuits, press well down and cover. Leave 24 hours to chill.

Of many tarts, gâteaux and biscuits that owe their flavour in part to a judicious portion of ground almonds, the Austrian *Linzertorte* is justly popular, and accordingly there are many variations of the recipe. One version is made with blanched almonds (not the following), another adds hazelnuts.

LINZERTORTE

Cut 5 oz (150 g) butter into pieces and add 5 oz (150 g) plain flour, then add the following: 5 oz (150 g) sugar, 5 oz (150 g) ground almonds, a pinch of nutmeg and a pinch of cinnamon, the yolks of 2 eggs, and the grated peel of half a lemon. Lastly, put in 2 teaspoons rum. When this has been worked into a dough, let it chill for an hour or two, then roll out a good two-thirds of the pastry and line a buttered flan case with it. Spread a good raspberry jam over it, and roll out the rest of the pastry into strips to make a lattice pattern. Brush lightly over with beaten egg, bake for half an hour in a moderate oven (350°F/180°C/mark 4), and dust with icing sugar when done.

A useful almond preparation is *pralin*. Put equal quantities of sugar and unblanched almonds in a pan over low heat until the sugar melts. Stir gently and leave to cook until a rich brown; allow to cool, then pound to a powder. *Praliné*, as a cake, is almond paste with a layer of praline butter cream in between and on top, sprinkled with almonds; *pralines* are sweets, simply almonds covered in a light coating of sugar and tinted in various colours.

NÈGRE EN CHEMISE

Have ready some strong black coffee, and pour 2–4 tablespoons over ½ lb (225 g) plain chocolate broken up from a block into pieces in a pan. Cook over a gentle heat until a thick cream forms, then beat in 2 oz (50 g) butter piece by piece. Flavour with 1 teaspoon rum. When cool, stir in 3 oz (75 g) *pralin*. Whip ½ pint (300 ml) double cream until it is just stiff, and fold this into the mixture. Turn it out into a greased bombe mould. After chilling for a few hours, turn out the bombe on to a flat dish and decorate with a piped collar of whipped cream round the base and flat chocolate rounds as a fringe. Serve with crème Chantilly and sweet biscuits.

MARZIPAN

Put ½ lb (225 g) granulated sugar in a saucepan, and add 7 tablespoons water; stir over a gentle heat until the sugar dissolves. Bring to the boil (at 240°F/120°C), remove the pan from the heat and add 6 oz (175 g) ground almonds and the white of an egg. Stir vigorously, return the pan to the heat to cook the egg white, then turn out the mixture on to a slab. Work it with a spatula while it cools and then knead in 1 tablespoon icing sugar until a smooth texture is achieved. Shape the marzipan as desired.

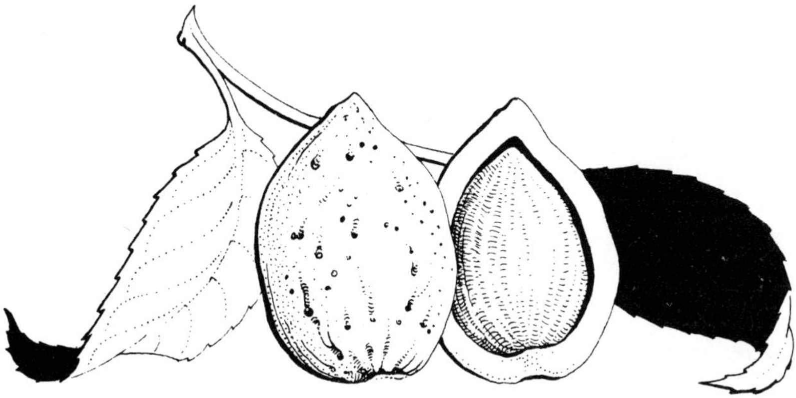

The following is just one of the many variations of the Spanish *turrones*, nougat-like almond sweets:

TURRON DE JIJONA

Blanch and skin 1 lb (450 g) almonds (see recipe for salted almonds), roast and chop them finely, then pound them in a mortar with ½ lb (225 g) sugar. Place the mixture in a saucepan with 4 oz (100 g) honey, and stir over the heat until the sweet begins to brown. It can then be poured into small rice-paper-lined frames or boxes until it sets. Cut into squares when luke warm.

Greek 'almond pears' are a dessert sweet – the recipe should yield about 24 shapes:

AMYGTHALOTA

Grind together 1 lb (450 g) blanched almonds with 2 tablespoons sugar until a very fine mixture is formed. Add a few drops of vanilla essence, the whites of 3 eggs, lightly whisked, 1 oz (25 g) soft breadcrumbs and 6 oz (175 g) sugar. Knead this to make a dough, and shape pieces of it into small pear shapes, inserting a clove in each to represent a stem. Place the almond pears on a buttered baking sheet and bake for 15 minutes in a moderate oven (350°F/180°C/mark 4). When cool, dip the pears in orange-flower water and dust heavily with icing sugar.

NAMES: Fr. *amande (douce ou amère)*; Ger. *Mandel*; Sp. *almendra*; It. *mandorla*; Latin, Gk. *amygdala*; Hebrew *shakod* ('the early').

STORAGE: Kept dry, packed in boxes or jars with tight-fitting lids, shelled almonds will keep for a long time.

Apple

JOHN GERARD, in 1597, stated nearly all that can be said about apples: 'Apples do differ in greatness, forme, colour and taste; some covered with a red skin, others yellowe or greene, varying infinitely according to the soyle and climate; some very great, some little, and many of a middle sort; some are sweet or tastie, or something sower; most be of a middle taste betweene sweete and sower, the which to distinguish I think it impossible.' He was writing at a time when the cultivated apple *(Pyrus malus)* was already of great antiquity and numberless in its varieties. Pliny describes the Romans as having twelve different apples, including the Spadonium, distinguished by the absence of pips. However, a coreless apple has eluded the efforts of all hybridizers; certainly it was not among those introduced from overseas into Britain during the Roman occupation.

In English history, the Pearmain is the first recorded variety with a name: it was cultivated in Norfolk in 1200 and remained in esteem until the eighteenth century. Next is the mediaeval Costard, popular in Shakespeare's time (and preserved in the word 'costermonger'), then follow the Pippins, of which the most famous dessert apple, the Golden Pippin, originated in Sussex. The Elizabethan age is rich in references to apples long merged into other varieties: Queenings, Leathercoats, Apple-johns, Pomewaters, Codlins, Quarrendens. The rise of horticulturalists and nurseries in the seventeenth century saw Redstreaks, Calvilles, Pommes d'Apis, and Nonpareils; the eighteenth century added the Ribston Pippin and, about the time of Waterloo, the Blenheim Orange. There then appeared Thomas Andrew Knight, a masterly fruit grower who raised, among other new fruit, twelve new varieties of apple. Horticulture owes much to his scientific genius, but his name is scarcely as well known as that of the retired brewer of Colnbrook, Mr Cox, who raised an Orange Pippin in 1830. Victorian apples are many. Bramley's Seedling, an excellent cooker, was raised by a Nottingham shoemaker; Worcester Pearmain, Newton Wonder, James Grieve and the Laxton apples were careful introductions into commercial apple-growing by professional gardeners. The names of a few other apples may be noted: Calville Blanche d'Hiver, a French variety that has long been considered the best of apples; the German Gravenstein; Egremont Russet, Newtown Pippin – but there is no end to this list of mere names, for it is claimed that there are some 7,000 named varieties of apple.

There must be as many apple dishes known to cooks all over the world. Dessert use apart, the apple is very amenable to cooking, and the three liquids it provides – apple juice, cider and apple brandy (applejack or *calvados*) – all have their roles. Apples appear in soups, stuffings and salads, with seafood, meat (pork and goose are esteemed partners), eggs and cheese, with some vegetables (e.g. potatoes and red cabbage), in sauces, relishes and glazes, and in all kinds of cakes, breads, pies, tarts, puddings, fritters, dumplings and other desserts.

For cooking purposes, Bramley or American Newtown, Granny

Smith, Sturmer and Cox's Orange Pippin are among the best. There is no need to narrow the choice to 'cooking apples'; dessert varieties, or mixtures of dessert and cooking, give the best flavour. Red-skinned apples, by and large, are perhaps better in salads and with cheese. Apples take kindly to rum and kirsch, to cinnamon and cloves, to butter and brown sugar. The apple is on the whole an untemperamental fruit, adaptable to a great variety of dishes.

That the apple has become a symbol of health is fortuitous (it contains a lot of fibre, and valuable minerals, but is by no means the most salutary of fruits), and is perhaps connected with the mass of legend and folklore it has attracted. Certainly in temperate climates the notion seems to prevail that if any one fruit is to represent all the fruits of the earth it is an apple in perfection, and as the first of fruits it offers endless culinary variety.

First, a Russian savoury dish, *yaichnitsa s yablokami*:

APPLES AND EGGS

Peel 1 large apple, core it, and cut it into 4 rings. Dip each of these in flour and fry both sides in butter, and while they are frying break an egg into each ring. Sprinkle with salt, and continue to fry until the eggs are set.

And a Polish dish:

VEAL AND APPLE CASSEROLE

Cut up into 1-inch (2½-cm) squares about 1½ lb (675 g) veal and brown in pork or goose fat. Remove, and use the same fat to fry 2 onions until golden brown, adding 4 moderate-sized cooking apples that have been pared and sliced, ¾ pint (450 ml) white stock, some ground cloves, and 1 teaspoon brown sugar to allay the tartness of the apples. Stew until smooth and pour over the veal in a casserole. Place in a moderate oven (350°F/180°C/mark 4) for 45 minutes to one hour. Serve with sour cream.

This Arabian dish consists of apples stuffed with chicken:

FOUJA DJEDJAD

Cut the meat off a cold, cooked chicken, and chop it up finely without mincing it. For every ½-pint (300-ml) cup that it fills, chop in 1 hard-boiled egg, and mix in salt and pepper and some sultanas. Take a number of firm apples, remove the cores, run the point of the knife round the skin a third of the way down to prevent bursting, and hollow them out, leaving about ½ inch (1 cm) of fruit next to the skin and over the base. Pour a little melted butter in, fill with the chopped chicken and egg, and cover each centre with ½ teaspoon honey. Place the apples in a baking tin in which 2 oz (50 g) butter has been melted, and cook in a slow oven (300°F/150°C/mark 2) for half an hour, or until apples are soft, basting all the time with butter.

LEICESTERSHIRE MEDLEY PIE

Leaving the peel on a number of apples, cut them into quarters and remove the cores. Cut substantial slices of cold boiled ham, and any other cold meat that has been roasted. Season these with a mixture of salt, pepper and ground ginger, then place in a baking dish to form alternate layers with the apples, cutting the apples up smaller if necessary to form even layers. Pour ½ pint (300 ml) beer over the top. Cover the dish with a suet pastry (see p. 12). Bake for at least half an hour in a moderate oven (375°F/190°C/mark 5).

GÄNSELEBER MIT ÄPFELN UND ZWIEBELN

Soak goose liver in milk for 1 hour, then dry, roll in salt and paprika, and finally in breadcrumbs before frying lightly in butter. Add 2 peeled and sliced apples and 1 large sliced onion. Fry until the liver is tender. A sauce made of madeira and truffles is sometimes made to accompany the liver.

These stuffed apples are served with pork chops:

SHROPSHIRE STUFFED APPLES

Peel and core 6 fair-sized cooking apples. Have ready half a ½-pint (300-ml) cup of stuffing composed of mixed chopped onion and sage, pepper and salt, and use it to fill the centre of each apple. Then put 2 teaspoons butter in a shallow pie dish, add 3 tablespoons water, and place the dish in the oven for 5 minutes to melt the butter and heat the water. Now place the stuffed apples in carefully, with butter on top of each, to bake (350°F / 180° C / mark 4) for 45 minutes to one hour, or until the apples are soft.

These with roast pork or roast goose:

ÄPFEL MIT LEBERFULLE

Prepare 8 large cooking apples by washing and hollowing out the centres. The stuffing is a mixture of 2 oz (50 g) pork, 1 goose liver, the apple centres less pips, and a sprinkling of marjoram and thyme, all of which should be passed through a mincer. Fill the apples with this, and arrange them in a buttered dish. Add 4 tablespoons white wine. Cover and bake for about 45 minutes to one hour in a moderate oven (350°F / 180°C / mark 4).

And these with hare, or any game:

POMME GASTRONOME

For each person, take 1 eating apple, as large as possible, and sever the top to form a round lid. Then carefully scoop out the pulp, removing the core and taking care not to pierce the skin from inside. Mix the pulp immediately with a few drops of lemon juice to prevent discolouration, then add a small amount of port or sherry, a little castor sugar, a pinch of nutmeg and a spoonful of cream. Fill the hollow apple with this mixture and put back the lid. Serve with the roast, and supply this side dish with its own spoon or fork.

'Heaven and earth' is simply a side vegetable dish of apples and potatoes to go with fried sausages, fried liver, etc:

HIMMEL UND ERDE

Cook 1 lb (450 g) peeled and quartered potatoes in sufficient salted water to half cover them until tender. Strain and mash. In another pan cook an equal quantity of prepared, tart cooking apples and heat to a smooth purée with 2 teaspoons sugar. Combine with the potatoes. Fry 2 oz (50 g) diced bacon until crisp, remove from the pan and, in the same fat, brown 1 large onion, thinly sliced. Add both to the potato and apple mixture. Season with salt and pepper. Serve alone or with fried liver or slices of black pudding.

APPLE DUMPLINGS

To prepare the crust, sift 5 oz (150 g) plain flour with a pinch of salt, and rub in 3 oz (75 g) mixed butter and lard. Stir in cold water, enough to form a stiff dough, then roll out into rounds large enough to envelop completely each of the 4 apples to be baked. Next core and peel the apples and position on each round of pastry. The apples should be stuffed with dried fruit, grated or candied lemon peel, and brown sugar. Draw the pastry up round the apples and join it securely. Brush over with beaten egg and dust with castor sugar. Place in a moderately hot oven (400°F/200°C/mark 6) and bake for 10 minutes, then continue to bake at a reduced heat (350°F/180°C/mark 4) for about half an hour, by which time the apples will be tender and the dumplings ready.

One of the many sorts of dumpling favoured in German regional cooking:

APFELKLÖSSE

Peel and core 1 lb (450 g) apples and cut into small pieces. Blanch 1 oz (25 g) sweet almonds and grind them finely. Beat 4 eggs into a froth, add 2 oz (50 g) melted butter, then mix in the apples and almonds and the remaining ingredients: 4 oz (100 g) currants, 4 oz

(100 g) soft breadcrumbs, 3 oz (75 g) sugar, $\frac{1}{2}$ pint (300 ml) milk, and the grated rind of half a lemon. If the resulting dumpling mixture is not firm enough, it can be stiffened with further breadcrumbs. Have ready some boiling salted water and drop into it small round dumpling balls made from the mixture. Cook for 15 minutes. Serve hot from the pan, basted with browned butter and sprinkled with brown sugar to which a pinch of cinnamon has been added.

APPLES À LA SUEDOISE

Dissolve $\frac{1}{2}$ lb (225 g) sugar in $\frac{1}{2}$ pint (300 ml) water to make a syrup, and into this grate the rind of a single large lemon. Then take 6 medium-sized dessert apples, peel, halve and core them, and cook them in the syrup in a covered saucepan until just tender. Arrange them in a dish when ready. Simmer to reduce the syrup until it is thick, adding a little lemon juice as this is done, and pour it over the apples. Let the dish stand overnight to set.

APPLE CHARLOTTE

Essentially a dish of bitter apples tempered by very thick cream; to this end, use only tart apples. Peel, core and slice 2 lb (1 kg) apples. Line the bottom of a buttered pie dish with them, sprinkle brown sugar over them, squeeze lemon juice on that and lastly cover with a layer of not too fresh white bread and butter. Repeat until the apples are used up, the top layer being thin bread and butter. Cover with greaseproof paper. Bake for 1 hour at a moderate heat (350°F/180°C/mark 4). Take the paper off, and let the crust brown. Serve very hot, with ample thick cream.

This is a Greek dish, *mila sto fourno me kastana*:

BAKED APPLES STUFFED WITH CHESTNUTS

Wash 6 large apples, and core them. Boil and peel between 24 and 36 chestnuts, break them into pieces without mincing them, then stuff them, together with 4 oz (100 g) large, stoned raisins, into the apple centres. Cover with a light sprinkling of sugar and cinnamon, and arrange the apples in a baking dish. Add 1 pint (600 ml) sugar syrup made from $\frac{1}{2}$ lb (225 g) sugar and $\frac{1}{2}$ pint (300 ml) water.

Bake in a moderate oven (350°F/180°C/mark 4); the apples are ready when they are soft – they will probably take about 45 minutes to one hour.

BANBURY APPLE PIE

Peel and slice 6 medium-sized apples and place a layer of them in a well-buttered pie dish, adding a sprinkling of chopped candied peel, a few currants, a little melted butter and a sprinkling of sugar. Repeat, layer by layer, until the dish is full: 4 oz (100 g) currants, 2 oz (50 g) peel and 3 oz (75 g) butter, melted and cooled, will suffice. Cover with shortcrust pastry (see p. 11). Glaze with milk and sugar and bake for 15 minutes in a moderately hot oven (400°F/200°C/mark 6) and then reduce the heat and bake for a further 15–20 minutes at 350°F/180°C/mark 4, or until the apple is cooked and the pastry is golden brown.

The name 'black cap' is usually applied in eighteenth-century cooking to a pudding in which currants drop to the bottom and then form a black top when the pudding is reversed out of its mould, but here to the burnt tops of the apples. This recipe from the same century is from *The Art of Cookery, by A Lady:*

BLACK CAPS

'Cut twelve large apples in halves, and take out the cores. Place them on a mazarine [a deep plate, or baking tin] as close together as they can lie, with the flat side downwards; squeeze a lemon in two spoonfuls of orange flower-water, and pour over them; shred some lemon-peel fine, and throw over them, and grate fine sugar [sprinkle castor sugar] all over. Set them in a quick oven, (425°F/220°C/mark 7) and half an hour will do them. When you send them to table, throw fine sugar all over the dish.'

This version of pandowdy, the old New England apple dish, was recorded by the novelist Edith Wharton:

PANDOWDY

'Peel and quarter enough apples to fill an earthern crock. Put in one pint [1 lb (450 g)] of sugar, one pint (600 ml) of water, and a cup

[10 oz (275 g)] of molasses. On top put one tablespoonful of ground cinnamon and nutmeg. Make a thick crust of bread dough over the top. Bake four hours [in a very slow oven (275°F/135°C/mark 1)]. Taste to see if right. Break the crust in small pieces into the crock, mix the crust with the apples, and let it stand in the oven an hour longer. It should be served cold, with cream.'

Pandowdy, sometimes called cobbler, is now usually made in a buttered dish, ground cloves are added to the ingredients, and it is more often a crusted pie than a form of crumble. Where the recipe refers to a bread dough, you could use a very thick layer of short-crust pastry (see p. 11), and cook for a much shorter time.

This Lebanese dish is alleged to have been introduced by Lady Hester Stanhope after she chose to live, on Pitt's death in 1806, among the Arab tribes of the Lebanon:

LADY STANHOPE'S BAALBEK APPLES

Peel, core and quarter 2 lb (1 kg) of good eating apples. Melt 4 oz (100 g) unsalted butter in a frying pan, and heat until it turns brown. Put in the apple quarters and brown them, turning with a flat knife. Use a high flame, removing the apples only when they are about to burn. Serve arranged on a dish and sprinkled with a mixture of castor sugar and cinnamon, about 1 tablespoon ground cinnamon to ½ lb (225 g) sugar.

The French classic apple tart, usually called *tarte Tatin*, comes from the Orléans area, the plain of Sologne, where the village of Lamotte on the river Beuvron lies on the southern road from Orléans to Vierzon. This is the original recipe, and the recommended apple is a red-streaked reinette – though it is said that the tart can equally be made with peaches.

TARTE DES DEMOISELLES TATIN DE LAMOTTE-BEUVRON

Take a copper-plated tin or baking tin about 2½ inches (6 cm) deep and line it thoroughly with butter. Next cover the bottom with castor sugar to a depth of ½ inch (1 cm), and fill the dish with a layer of sliced apples. Cover the apple slices with a little butter and sugar, then place over the whole a thin layer of *pâte brisée* (see p. 11). Cook over a good charcoal fire in a salamander for 20–25 minutes; check by lifting the edge of the *pâte* with a knife – the tart is ready when the apple is golden and the sugar starting to caramelize. At this stage, cover the tart with the plate on which it is to be served and quickly turn the whole tart over so that the apples are upwards. Serve hot.

Ordinary oven cooking for 30 minutes at a moderate heat (350°F/180°C/mark 4) should suffice in place of the more complicated *four de campagne* specified.

POMMES NORMANDES

Cut up some eating apples, preferably Cox's, into quarters, and then cut each quarter into 6 pieces. Dry well. Melt some butter in a pan, put in the apples and fry them on all sides, adding sugar. When cooked, add a little more sugar and a very little ground cinnamon. When ready to serve, sprinkle on some kirsch and light it. Serve whipped cream flavoured with kirsch with the apples.

SWEDISH APPLE PURÉE

Wash and dry $1\frac{1}{2}$ lb (675 g) sour apples, core them, cut into sections, and cook in a little boiling water to which the finely peeled rind of a quarter lemon and a small piece of cinnamon have been added. When tender, pass the apples through a sieve and heat this purée with 4 oz (100 g) sugar. When boiling, add up to 2 tablespoons potato flour mixed with 5 tablespoons cold water. Add 3 tablespoons white wine and any more sugar that seems necessary. Serve cold with whipped cream.

SWEDISH APPLE PUDDING

Peel and core 2½ lb (1¼ kg) apples, and slice them into thin segments. Cook in very little water with 4 oz (100 g) sugar and reduce to a pulp. Grate half a stale brown loaf – enough to make about 6 oz (175 g) breadcrumbs – and brown lightly in 3 oz (75 g) butter. Butter a tin, and fill it with alternate layers of breadcrumbs and apple, three of breadcrumbs and two of apple, the top and bottom layers being breadcrumbs. Dot the top with spots of butter, and bake for half an hour in a moderate oven (350°F/180°C/mark 4). Turn out, and sprinkle with castor sugar. Serve hot or cold, with whipped cream.

STEWED APPLES

Peel, core and cut up 1 lb (450 g) juicy eating apples. Put them in a large saucepan with ¼ pint (150 ml) water and 2 tablespoons sugar, the rind of half a lemon, and 1 tablespoon rum. Stew gently until the apples fall into a purée. When they are cold, stir in 2 oz (50 g) chopped, candied peel. Before serving, add a further tablespoon of rum, and have whipped cream ready.

APPLE GINGERBREAD

Peel ½ lb (225 g) cooking apples, slice them, and place in a pan with 1 dessertspoon Demerara sugar and a little water to prevent them burning. Stew until tender, mash, and leave to cool. Put 4 oz (100 g) golden syrup in a pan with 3 oz (75 g) butter and 2 oz (50 g) Demerara sugar, dissolve slowly, and leave to cool. Mix 6 oz (175 g) self-raising flour in a basin with 1 teaspoon ground ginger and ¼ teaspoon ground cloves. Whisk up an egg, add the dissolved syrup, and add this to the flour. Mix well, then stir in the mashed apple and beat it all together. Turn all the mixture into a well-buttered oblong tin, and bake for 40–45 minutes in a moderate oven (350°F/180°C/mark 4). Let it stand a while in the tin before turning out. A cinnamon icing can be added, made of 6 oz (175 g) icing sugar mixed with 1 teaspoon ground cinnamon and a little hot water – it will set on top of the gingerbread if evenly spread.

NAMES: Fr. *pomme*; Ger. *Apfel*; It. *poma*; Sp. *manzana*; Dutch *appel*; Swedish *appel*; Russian *yabloko*.

STORAGE: Unripe apples can be stored until ripe, and ripe fruit, in a cool place where they will not dry out, will last some weeks before the skin starts to shrivel.

Apricot

A STURDY tree, *Prunus armeniaca* flowers early in the year and yields late summer fruits of varying colour: pale yellow, orange, yellow-orange with red-freckled skin. In a temperate climate, late frosts are an enemy to the fruit; only the warmer south-west walls of Britain will accommodate the tree. It is generally said to have been brought to England in 1524, the end of a long journey from North China, probably Tien-Shan, where it is native – not from the Caucasus as its name suggests, although it had long been cultivated in Armenia. The apricot was spread by the Arabs, who retain the fruit in Middle Eastern cookery because of its compatibility with lamb. Japan, North Africa, California, the Cape of Good Hope and the Mediterranean are all places where the apricot flourishes; the best fruits are deemed to be those of the Loire valley. The apricot grows wild in Siberia and in Afghanistan around Kabul, where, additionally, there is claimed to be a variety of great delicacy and lusciousness.

The fruit itself varies considerably in quality; dried apricots are perhaps more versatile than most dried fruits, retaining a better flavour than many fresh apricots. A dull-coated fruit or a greenish one is either over-ripe or under-ripe, and a sound appearance may conceal the mealy texture of flesh that has ripened on some long journey from overseas. A perfect fruit, of course, should be eaten as dessert, and not cooked; it has been suggested that only a glass of Château Yquem could be a fitting accompaniment.

In 1548, when the apricot was still a rare fruit, William Turner wished it to be named 'an hasty peach tree' and Dr Muffet gave it as his opinion that 'Apricots are Plums concealed beneath a Peach's coat', but its individual nature and its own peculiarly rich flavour have long since earned the apricot its devotees. This may be due to the high potassium level and the fact that the apricot has one hundred times the average vitamin A found in other fruits. The kernels within apricot stones, although cumulatively poisonous, are the source of the excellent *eau de noyaux* liqueur. An apricot wine is sometimes made in England, and in France in particular crystallized fruit, preserves, confectionery, dried pastes (a *pâte* d'Auvergne is famous) and jam embody much of the produce.

Few admirers of the apricot ever tire of the fruit. When Ude, the great chef, was maître d'hotel at Crockford's, he let rooms to Lord Alvanley, the Regency epicure, and every day throughout the year supplied him with a cold apricot tart. Ude left no recipe in his cookery book, so perhaps this was the simplest of items, indispensable by virtue of using one of the pleasantest of fruits.

APRICOT SOUP

Cover 1 lb (450 g) apricots with water and let them cook very slowly. Pass through a sieve, with the juice. Add both water and red wine in sufficient amounts to make the desired quantity of soup, add a little sugar to sweeten, some arrowroot mixed with a little cold water to thicken, and a seasoning of cinnamon. Serve very hot, with diced fried bread. Sour cream may also be served with apricot soup.

GAMMON AND APRICOT PIE

Take a large gammon rasher, which should be 1 inch ($2\frac{1}{2}$ cm) thick, and brown it lightly in butter on both sides in a frying pan. Set it in a pie dish. Arrange $\frac{1}{2}$ lb (225 g) dried apricots, which have been soaking in cold water for at least 12 hours previously, on top of it. Grind some pepper over the fruit, sprinkle on 1 oz (25 g) sultanas, and pour some brown stock over the dish. Cover with sliced potato – 6 potatoes would suffice – and place greaseproof paper over the pie dish. Bake in a moderate oven (350°F/180°C/mark 4) for 1 hour, and serve hot.

The Arabic for apricot is *mishmish*, giving the name to the meat and apricot dish. It is intended for mutton and particularly makes use of the fat from the fat-tailed oriental sheep. The tail is reduced and the meat, cut up small, is browned in the dripping.

MISHMISHIYA

Cut up fat meat into small pieces, and brown in dripping, stirring constantly. Cover the pieces with water, and add some salt and a little cinnamon. Boil the meat, skim off the scum, and throw in onions on top of the meat, adding finely ground coriander, cumin, mastic, cinnamon and ginger. Soak dried apricots in water, boil, and then strain the juice; add it to the saucepan. Grind sweet almonds, mix with water, colour with saffron and add to the saucepan. Sprinkle rose water over the saucepan, and wipe the sides clean. Leave the dish to settle over the fire for an hour, then remove and serve, with the apricots and rice.

The details vary in the different versions of the recipe: the basic amounts can probably be equated with 2 lb (1 kg) lamb, 2 onions and ½ lb (225 g) dried apricots; 2 oz (50 g) almonds would suffice, while ¼–½ teaspoon each of the spices, according to taste, would be ample; 1 teaspoon rose water is required. The total cooking time should be about 2 hours, cooking very slowly. A mound of rice should be prepared separately.

Apricot

Both Austria and Germany have recipes for dumplings containing apricots:

MARILLENKNÖDEL

Ease the stones out of 16 ripe apricots (for 4 people), and in their place put a cube of sugar. Prepare a dumpling paste as follows: boil $\frac{3}{4}$ pint (450 ml) water, slightly salted and with 1 oz (25 g) of butter in it, and stir in $\frac{1}{2}$ lb (225 g) flour until the dough no longer adheres to the sides of the pan. Remove from the heat, and beat in 2 eggs to make a smooth paste, leaving this to cool in a bowl. Knead the dough when cold. Cover each apricot completely, drop into a large pan of boiling water, cover and cook; as the dumplings rise when ready leave them to simmer for a further 2 minutes. Lift out with care, rinse in cold water in a sieve, then fry in a saucepan in 2 oz (50 g) butter with a good measure of fresh breadcrumbs that have already been browned. Toss the dumplings, and serve sprinkled with castor sugar.

This standard recipe for almost any fruit pudding, based on the recipe in *Adam's Luxury and Eve's Cookery*, 1744, can be used for apricots:

BAKED APRICOT PUDDING

Steam 6 large apricots until they are very tender. Break them up into small pieces and sweeten with sugar to taste. Let them grow cold, then beat in the yolks of 4 eggs, the whites of 2 eggs and a little cream. Butter a pie dish and line it with puff pastry (see p. 12), put the mixture in it, and bake – for 10–15 minutes in a hot oven (425°F/220°C/mark 7) to do the pastry, and then for 15 minutes at a moderate heat (350°F/180°C/mark 4) until the fruit mixture sets firm.

BAKED APRICOTS

Cut in half 1 lb (450 g) apricots and stone them. Lay them in a fireproof dish and cover with brown sugar. Leave for 12 hours, by which time the juice will have exuded. Bake in a moderate oven

(350°F/180°C/mark 4) for 20 minutes, and leave to cool. Serve cold with whipped cream.

A Spanish cold dessert:

ESPUMA DE ALBARICOQUE

Peel and stone 2 lb (1 kg) apricots, and cut into slices. Put them in enough water to cover them, and cook until tender. When cooked pass the fruit through a sieve with some of the juice. Use the rest of the juice, and 1 lb (450 g) sugar, to make a syrup, and mix this with the apricot purée. Allow to cool, then fold in the whites of 3 eggs, stiffly beaten. Mix well, and freeze; before serving, sprinkle with a few finely chopped almonds.

TARTE AUX ABRICOTS

Make a pastry with ½ lb (225 g) flour and 4 oz (100 g) butter, an egg and a pinch of salt; knead with a little water and leave for 1 hour. Roll out the pastry and use to line a buttered tart tin. On top of the pastry arrange apricots that have been halved and stoned, cut sides up. Drop a little sugar into the centre of each apricot. The fruit should be packed tightly to allow for shrinkage in cooking. Put in a moderate oven (350°F/180°C/mark 4) to bake for 40–45 minutes, or until the apricots are soft and the pastry golden brown. Take a few extra apricots and stew them with sugar, reducing the liquid to make a syrup and adding a few drops of kirsch; when this is cool use to pour over the fruit to form a glaze. Serve with whipped cream.

NAMES: Fr. *abricot*; Ger. *Aprikose (Marille*, Austria*)*; Sp. *albaricoque*; Port. *albricoque* (= Elizabethan Eng. 'apricock'); It. *albicocca*; Gk. *mela armeniaca.*

STORAGE: In a cool place, with some humidity, apricots will keep up to four or five days.

Avocado

A HUNDRED years ago, the scholarly authors of *A Glossary of Anglo-Indian Words* commented on the recent introduction of the avocado to Bengal that 'had it been worth eating it would have come long before ... with liberal pepper and salt there may be a remote suggestion of marrow: but that is all.' And in 1919 Wilson Popenoe, the distinguished agricultural explorer, was wondering whether the avocado, virtually the meat of the Central Americans, would become a staple food in North America. Tastes change, and yet the versatility of the fruit is still far from wholly exploited. For centuries the avocado had been cultivated and enjoyed in its native home, the American tropics; the name itself is a version of the Aztec *ahuacatl*, and there are many similar corruptions of the name. The early Spanish chroniclers of Central America are the first Europeans to mention the fruit, and its common name of 'avocado or alligator pear-tree' appears in Sir Hans Sloane's catalogue of Jamaican plants in 1692, the fruit having been planted there by the Spaniards. Avocados are now cultivated in most tropical areas, Australasia, South Africa, the West Indies, California and Florida in America, Hawaii, the Philippines, the Mediterranean countries and Israel, as well as the original localities in South America.

The avocado belongs to the genus Persea and hence to the laurel family, and is so related to cinnamon, camphor and sassafras. There are three distinct races: (1) the Mexican, thin-skinned *(Persea drymifolia)*, (2) the West Indian, thicker-skinned, and (3) the Guatemalan, with a thick, woody skin; both the latter are *Persea americana*. From these a number of varieties and hybrids have been established, and of these four kinds are worth knowing by name. The Fuerte is pear-shaped, dark green with many small yellow dots; a hybrid of the Mexican and Guatemalan types, it originated in Puebla, Mexico, in 1911. The Nabal (Trapp, *P. americana leiogyna* in America) is large and round, with a large stone. The Ettinger is long, pyriform, green and smooth-skinned. The Hass, which bears some relation to another propagation of 1911, then known as the Puebla, has a distinctive ovoid shape and a rough purple skin which turns black when ripe. Its stone is small, and its flesh smooth; it is a preferred type. All these, however, can be used in the same way, and little knowledge of the behaviour of the fruit is required. Ripeness is important, as the unripe fruit is inedible; the first thing you need to know is that the fruit does not ripen on the tree, and would rot if not gathered. Ripeness is not necessarily apparent by any change of colour in the skin, so the fruit must be felt; if it yields slightly all over it is ripe. Discoloration of the flesh of cut avocados can be avoided by sprinkling with lemon juice.

Nourishing as a fruit, the avocado contains much protein, vitamins A and B, and a percentage of oil, about 25 per cent, varying from type to type. Old names like Midshipman's Butter and Butter Pear suggest its value as a spread. Mashed and seasoned with vinegar, served with chopped onions, pickles and pepper as a filling, or simply eaten raw with mayonnaise or oil and vinegar, the avocado suggests a large number of uses and dressings. South Americans dice the flesh and add it to hot soup at the time of serving. It will mix with lobster, crab, chopped dates, ginger, fruit juices, and sauces and seasonings of many kinds. The *guacamole* of Mexico (see p. 243), with tortillas, is a substantial dish. A vinaigrette dressing with black pepper is a basic dressing that well suits the buttery flesh, but fillings, spreads, seasonings and dressings are all a matter of experiment and preference.

AVOCADO SOUP

Open 2 ripe avocados and remove the stones; put the flesh in a bowl and mash with the yolk of 1 egg, then add 3 tablespoons cream. Bring $1\frac{1}{2}$ pints (900 ml) chicken stock to the boil, then remove from the heat and whisk in the avocado mixture. Blend vigorously, add salt and pepper to taste, then set aside to cool. Refrigerate for 1 hour before serving.

AVOCADO À LA RUSSE

Cut each avocado pear in half lengthwise, and brush the flesh with lemon juice at once to avoid discolouring. Scoop out the flesh, cut up and replace it in the skin, leaving a cavity in the middle which is then filled with caviare. Serve chilled.

AGUACATES RELLENOS DE PICADILLO

Peel 2 medium-sized avocados, cut horizontally, remove the stones and brush the flesh at once with lemon juice to avoid discolouring. Mix a stuffing of 4 tablespoons minced cooked pork, 2 small chopped tomatoes, 6 capers and ½ teaspoon chopped marjoram. Season the avocado halves with salt and pepper and stuff them with the mixture. Have a beaten egg ready, dip the halves in it, then cover with fresh breadcrumbs; fry in butter until completely browned. Then lay the avocados in a baking dish, cover with thinly sliced tomatoes or a home-made tomato sauce, and cook for 10 minutes in a hot oven (425°F/220°C/mark 7). Serve hot.

In Nicaragua, this South American recipe is varied by using a stuffing of fresh cream cheese – a curd cheese, but any soft cream cheese would be suitable.

AVOCADO AND TUNA FISH

Remove the stones of 3 ripe avocados and mash the flesh, blending in first the juice of 1 lemon, then 3 tablespoons olive oil, half a mild onion, grated, some chopped parsley, salt and white pepper to taste, and a crushed garlic clove. Smooth the purée in a blender. Drain all the oil from a tin of tuna fish, and combine the flakes of fish with the purée. Serve on thin, warm toast.

AUSTRALIAN AVOCADO SALAD

Take 3 large, ripe avocados and cut in half lengthwise. Remove the stones. Scoop out the flesh and leave the shells intact, brushing the insides with lemon juice. Cut the flesh into ½-inch (1-cm) cubes and combine with the flesh of 1 small fresh pineapple, chopped to the same size, 4 oz (100 g) small ripe strawberries and 2 oranges, peeled and diced. Prepare a dressing for the fruit by mixing 5 tablespoons olive oil, 5 tablespoons pineapple juice, 1 tablespoon lemon juice and ½ teaspoon salt. Pour this over the fruit, chill for 1 hour, drain off the dressing, and fill the shells with the fruit mixture. The surplus dressing can be blended with mayonnaise and served separately.

A French colonial recipe from Madagascar:

AVOCAT MALGACHE

Halve the fruit and remove the stone. Into the two cavities put a filling that has been prepared overnight, consisting of grapefruit flesh soaked in a marinade of rum and sugar. Serve the avocado well chilled, topped with a pair of glacé cherries.

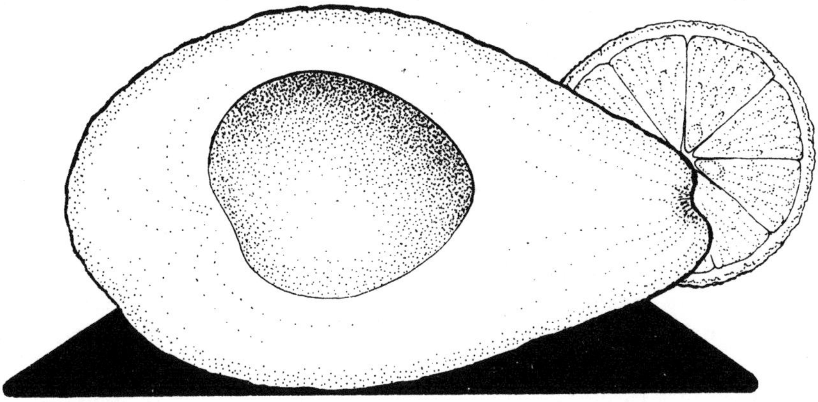

AVOCADO AND PINEAPPLE DELIGHT

Peel a soft avocado and mash the flesh with a fork until a smooth purée is achieved. Add 2 tablespoons fresh pineapple juice, 1 teaspoon vanilla essence and 2 well-beaten eggs. Blend all these together with a rotary beater. Whip $\frac{1}{2}$ pint (300 ml) double cream until it is thick, then fold in the avocado mixture, adding sugar if desired. Freeze the mixture in a tray until firm, and serve on chilled, drained pineapple rings.

AVOCADO CREAM

Beat 2 eggs lightly and whisk into 1 pint (600 ml) cold milk. Add 4 oz (100 g) sugar and cook in a double boiler until the mixture thickens. Add some grated lemon rind for flavouring and let the custard cool. Take 2 ripe avocados, remove the stones, and mash the flesh into a smooth purée. Add another 4 oz (100 g) sugar and the juice of 1 lime to the pulp and beat it into the custard. Chill thoroughly in a refrigerator.

NAMES: Fr. *avocat*; Sp. *aguacate*; Brazilian *abacate*; Peruvian *palta*; Dutch *advocaat.*

STORAGE: Avocados should be used if they are ripe. Once cut, they need the protection of lemon juice to prevent discolouring, and should not be refrigerated for long.

Banana

IN the Western world, the banana (the common banana, *Musa sapientum*) is a modern fruit, and arguably one of the most popular, establishing itself in the eating habits of most countries with ease and rapidity. Yet references to banana cultivation in India go back to 500 BC, in China to 200 AD, and the travels of the banana tree can be traced to both the Mediterranean and Polynesia a thousand years ago. By the early fifteenth century it had spread from Africa to the Canary Islands, and thence it was taken to Santo Domingo in the New World, by Friar Tomas de Berlanga, in 1516. In the City of London, a bunch of bananas was displayed in Johnson's 'herbalry' in the year 1627, but little more is heard of the fruit until the first bananas reached New York from Cuba in 1804, and even then the first whole shipload did not arrive at Boston from Jamaica until 1870. In 1878 Fyffe, Hudson and Co. shipped bananas in crates lined with cotton wool to England from the Canaries, but their importation in quantities sufficient for them to be regarded as a new food is due to the great shipping entrepreneur Sir Alfred Jones. His Elder Dempster line carried bananas in whatever spare space the ships had; the difficulty of getting the fruit to arrive in Britain at the correct stage of ripeness meant that he had to hire costermongers to load barrows from the ships as they arrived and sell them at once. Thus Liverpool and Lancashire were the first places to acquire a taste for bananas.

Two events furthered the trade in bananas: a sudden decline in the cochineal imports from the Canaries in about 1900, for which a substitute was necessary, and the introduction of refrigerated ships in 1901. The arrival of bananas was always news: *The Times* reported a bunch coming in from Madeira in 1848, and in 1905 the *Daily Telegraph* signalled the arrival of some 14,000 bunches, a theme again taken up by the press after the total absence of the bananas during the Second World War.

Why this avidity for bananas, which are repugnant to some Western palates? Victor Jacquemart, the naturalist, traveller and friend of Stendhal, wrote from India in 1829 to say that bananas 'resemble jasmine-scented pomade gone rancid and much sweetened'. But they have been the sustenance of native races for centuries. An English authority on fruits went so far as to write, not many years ago: 'The banana was designed as the staple food of the negroid races, and they eat it ... with the Calvinistic resignation that a Scotsman shows to porridge.' Very roughly, half the world's bananas are eaten raw when ripe, practically half are cooked, and the negligible remainder have minor uses, e.g. in beer-making (the *mbidde* of Uganda) and in other fermented drinks. The banana contains a valuable amount of vitamin A, rich amounts of B and C, minerals (iron, potassium, etc.), and a high sugar content in easily assimilable form. The virtue of the banana lies in the facility with which it is digested, which makes it an ideal food for the very young and for invalids.

There are of course many species and types of banana (green cooking bananas are true plantains, and they can be considered as vegetables), but for us the practical scope of dessert bananas is limited to the few kinds that can be exported and survive. Banana-producing countries have delicate varieties not suited to export; the short 'Lady's Fingers' variety and the red-skinned banana are kinds not seen in countries that import.

Apart from being eaten raw, bananas can be mashed with cream and rum, baked or flambéed with brown sugar, chopped into fruit salads, mixed into ice cream, cooked in tarts, pies and puddings, or combined with fish or meat, where their digestive qualities are an advantage. In the Caribbean, bananas and plantains dominate the culinary scene, Jamaica in particular making full use of the fruit in every commercial way. While there is no dish in *la grande cuisine* to which the banana is admitted, it could well be made more use of in day-to-day cooking.

ICED BANANA SOUP

Peel 4 ripe bananas, and mash them to a smooth pulp. Rub this through a sieve, slowly working in 2 pints (1·2 litres) cold milk and adding a pinch of salt and a little sugar. Grate into it the rind of 1 orange, and bring to the boil. Stir in a very little cornflour, enough to thicken the soup, mixed with milk; do not over-thicken. Leave the soup to get cold, then cool further in a refrigerator. Before serving beat the soup thoroughly, and add a few ripe, stoned cherries.

TURBOT WITH BANANAS

Fillets of turbot are required for this dish, cut along and not across the fish, to avoid any bones. Sprinkle each fillet with melted butter, and grill. At the same time, melt some butter in a frying pan and gently fry halves of bananas, cut lengthwise. When the fillets are done, serve them with a half banana on each, and with a *sauce Robert*, prepared earlier and heated up. *Sauce Robert* consists of chopped onions, fried in butter, white wine and stock, reduced to a suitable consistency, with a little mustard added just before serving.

WEST INDIAN PORK CHOPS AND BANANAS

Take 6 large bananas, peel them, and sauté lightly in butter. Form a purée by forcing them through a sieve and mixing with a pinch of ground nutmeg, a pinch of ground cinnamon and a few drops of rum. Use this to line an oven-proof dish. Now sauté 4 pork chops in butter until they are brown and cooked through. Lay them on the banana purée, reheat in the oven, and before serving flambé the whole with 2 tablespoons warmed rum.

BANANA LOAF

Cream together 4 oz (100 g) butter and 6 oz (175 g) brown sugar. Add 2 well-beaten eggs. Mash 3 ripe bananas and pass them through a sieve, then add 1 teaspoon lemon juice. Blend this with the butter mixture. Next, sift together 1 heaped teaspoon baking powder, 1 level teaspoon bicarbonate of soda, a pinch of salt and

10 oz (275 g) plain flour; stir these quickly into the banana mixture and add 2 oz (50 g) mixed chopped nuts. Turn the whole mixture into a buttered loaf tin. Bake in a moderate oven (350°F/180°C/mark 4) for 1–1½ hours to make a 1 lb (450 g) loaf.

BANANA TART

Make rich shortcrust pastry from 1 lb (450 g) butter, 1 lb (450 g) flour and proportionate amounts of salt and 2 teaspoons sugar (see p. 11). Roll on a floured board, trim to the desired shape, and build up the edges with strips of pastry; prick the bottom, chill, then line with wax paper and fill with rice to retain the shape. Brush the edge strips with a mixture of egg and 1 tablespoon cream and bake in a hot oven (425°F/220°C/mark 7) for 15 minutes. Dispose of the paper and rice, and bake for 10 minutes more. Let the shell cool, then cover with caramel made from 2 oz (50 g) sugar and 2 tablespoons water, cooked together until it turns a rich golden brown. Cut as many bananas as are needed into ½-inch (1 cm) rounds, and poach them in a rum syrup made from rum, brown sugar and water for a few minutes. Cover the pastry shell with almond-flavoured *crème pâtissière*. Cover this with the banana slices, as close together as possible, and brush them with an apricot glaze, made from sieved apricot jam heated with a little water.

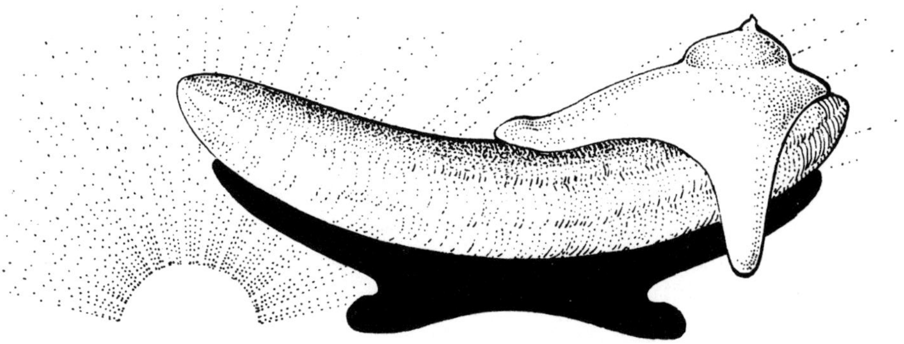

BANANA AND RAISIN PUDDING

Mash 4 ripe bananas and mix with 4 oz (100 g) stoned and halved raisins. Fold in 3 oz (75 g) brown sugar, and 6 oz (175 g) fresh white breadcrumbs. Beat 2 eggs, and add. Stir in 1 tablespoon lemon juice and a few shreds of grated lemon rind. Turn this mixture, thoroughly stirred, into a buttered mould, and cover with buttered greaseproof paper or foil. Steam for a full 2½ hours. The pudding can be served with cream or with an apricot sauce (see p. 243).

BUNDABERG BANANA FRITTERS

Peel 4 large ripe bananas and halve them lengthwise, place in a shallow dish and pour over a mixture of 2 tablespoons brown sugar, 2 teaspoons lemon juice, ½ teaspoon grated orange peel and 3 tablespoons rum. Let this stand for an hour, occasionally turning the bananas. Make a batter of 4 oz (100 g) plain flour, ¼ pint (150 ml) water and 1 tablespoon salad oil, then fold in a stiffly beaten egg white. Heat 5 tablespoons frying oil in a pan, drain the bananas, dip them in the batter and fry until golden brown all over. Heat the remaining rum mixture up as a sauce, first adding the juice of a fresh orange.

BANANES FLAMBÉS

Take 4 long, ripe bananas and, without peeling them, prod them with a fork in a few places; bake for about 10 minutes in a moderate oven (350°F/180°C/mark 4). Remove from the oven, slit the darkened skins horizontally and peel off the top half of the skin only. Melt 2 oz (50 g) butter in a frying pan, add 3 oz (75 g) Demerara sugar and ½ teaspoon allspice, and mix. Keeping the pan over a low heat, lay the bananas in this, skin uppermost, then peel the remaining skin off. Sprinkle fresh lime juice over the fruit. Warm 3 tablespoons rum in a small pan, light it, pour over the bananas and agitate the frying pan. Lift out the bananas to serve, with some of the sauce from the frying pan.

This Indonesian dish of baked bananas was imported from the Dutch East Indies, and is now part of Dutch cuisine:

PISANG GORENG

Peel 4 bananas and cut in half lengthwise. Put them in a well-buttered baking dish, dot with butter and sprinkle with salt and a liberal amount of lemon juice. Place in a hot oven (425°F/220°C/mark 7) and bake for 6 minutes.

CRÊPES SOPHIE VICTOR

Make a batter from 4 oz (100 g) flour, a pinch of salt, 1 egg and ½ pint (300 ml) milk or milk and water. Add the egg to the dry ingredients, then gradually add the liquid, beating until smooth. Cook 12 thin pancakes and set them aside. Peel 6 ripe bananas and slowly fry in butter – one at a time. As they are cooked, cut them in half lengthwise, sprinkle with castor sugar and wrap each half in a pancake, turning in the ends as you roll. Set in a buttered chafing dish to keep warm, and when serving pour over them a sauce made of butter, sugar and sufficient curaçao and Grand Marnier. This is then lighted at the table.

NAMES: Fr. *banane*, *figue d'Adam*, *figue du Paradis*; Dutch *banaan*; Ger. *banane*; It. *fico d'Adamo*, etc.; 'banana' is common to several African languages (but Swahili *ndizi* and over twenty names for different parts and varieties); Malay *pisang* (perhaps from Sanskrit *picanga* = brownish yellow); Chinese *kan chiao*.

STORAGE: Bananas should be used within a few days if ripe, and should not be placed in the refrigerator. For general cooking, slightly unripe fruit, with an inch (2½ cm) of green at the stalk end, is preferable; for eating, wholly yellow skins lightly spotted in brown or black are best. Bruised and blackened patches can always be cut out.

Berries

INNUMERABLE wild berries grow in the temperate and northern regions of the world; they were the first foods picked by man and, as the art of cooking has progressed, they have found their niche in providing agreeable but astringent dishes such as jellies, sauces, pies, puddings and wholesome syrups and cordials. There is also a huge range of berries to be found in tropical and sub-tropical regions, from the Ceylonese jambol *(Acronyctia laurifolia)* to the Mexican *Zizyphus obtusifolia*, but these are only eaten locally; acidulous, or without flavour, the majority have little to recommend them. It is in the woods, forests and hedges of northern Europe and North America that the most useful berries are found, and in practice these are limited to a few sorts, some of them found in several types. Many wild berries are just not eaten, and some few are inedible or toxic.

Barberry *Berberis vulgaris* is an attractive bush with yellow flowers and pendulous, elongated bright red fruits. Once candied in whole bunches as barberry drops, the acid berries can be used for a scarlet jelly, which traditionally goes with mutton pies, for a pleasant jam, and for pickling.

BARBERRY JELLY

Pick the barberries from their stalks, and bruise slightly. Cover with water in a preserving pan, boil up, and simmer until the fruit is soft. Place in a jelly bag to let the juice drip into a pan for at least 12 hours. Measure the collected juice, and allow 1 lb (450 g) sugar to every pint (600 ml) juice. Return to the preserving pan, let the sugar dissolve over a low heat, then boil up the liquid quickly until it jellies when tested (see jam section, p. 145). Put into hot jars and cover when cold.

Bilberry *Vaccinium myrtillus*, a widespread European berry found on moors and in woodland copses, with flesh-coloured flowers, and fruit of damson blue with a misty grey bloom. In England a number of local names occur: whortleberries, hurts, whinberries, or blaeberries. The bilberry is known in America and the name is given to several varieties, e.g. the dwarf bilberry *(V. caespitosum)* and the Kamchatka bilberry *(V. praestans)*. The usual dessert uses apply. In Germany, the berries *(Preisselbeere)* are used to make a sauce for game, and a wine is made from the fruit. Jams, jellies, fritters and tarts are appropriate to the sweetish, astringent taste – indeed, bilberry tarts, with baked apple supporting the berries in the tart, are claimed to be among the best of pastries. (Fr. *airelle*; It. *mortella*; Danish *bölle-baer*).

BILBERRY FRITTERS

Make sufficient fritter batter, unsweetened (see recipe for banana fritters, p. 44). Add bilberries until the batter is quite thick, then drop fritters into very hot oil or butter and fry on both sides until golden. Remove when ready and drain, then coat generously with castor sugar. Serve at once.

BILBERRY PIE

Mix 1 lb (450 g) bilberries with 3 oz (75 g) sugar. Take 2 cooking apples, bake, and remove the soft pulp. Mix with the bilberries, and sweeten. Butter a pie plate, cover it with puff pastry, (see p. 12) and fill with the bilberry mixture. Cover with a thin sheet of pastry, and brush this with egg white. Dust with castor sugar and bake first in a quick oven (425°F/220°C/mark 7) for 15–20 minutes and then in a slow oven (300°F/150°C/mark 2) to cook the fruit. Serve hot or cold, with cream.

NORTHUMBERLAND BLAEBERRY TART

Pick over the blaeberries – 1¼ lb (550 g) are needed – and wash clean; wash and wipe 1 lb (450 g) rhubarb and cut it into 1-inch

($2\frac{1}{2}$ cm) lengths; peel and quarter 1 lb (450 g) apples, remove the cores and cut into small pieces. Put an eggcup in the centre of a large pie dish, and fill the dish to the brim with the fruit, in layers. Use $\frac{1}{2}$ lb (225 g) sugar to sprinkle all over the fruit, add 2 tablespoons of water, and cover the dish with shortcrust pastry (see p. 11). Bake for 10 minutes at 400°F/200°C/mark 6 and then for 20 minutes at 350°F/180°C/mark 4.

Blueberry The great North American berry *(V. pensylvanicum)*, with many varieties (e.g. the high, or swamp, blueberry *(V. corymbosum)* inhabiting moist woods and bogs from Maine to Minnesota; the low, or sweet, blueberry *(V. angustifolium)* found on heaths and mountain lands from Newfoundland down to Virginia, and the small Canadian, or sour-top, blueberry). The large blue-black berries, sweet and agreeable to eat, were much used by the Indians. The berry is now cultivated commercially as well as growing wild, and its principal use is in one version or another of the traditional American blueberry pie, or in similar desserts.

BLUEBERRY PIE

Take a deep pie pan, and line with shortcrust pastry (see p. 11). Mix 1 lb (450 g) selected blueberries with 2 heaped tablespoons flour, 4 oz (100 g) sugar, and $\frac{1}{4}$ teaspoon salt, and place in the pan, adding more blueberries, flour and sugar in the same proportions if needed to fill the pan. Cover with a pastry lid, and bake in a moderate oven (350°F/180°C/mark 4) for about 45 minutes, until the crust is browned. Serve hot or cold, with cream.

STEWED BLUEBERRIES

To $1\frac{1}{2}$ lb (675 g) selected and washed blueberries, add $1\frac{1}{2}$ pints (900 ml) water, and sugar to taste, and bring to the boil. As the berries start to boil, add small dumplings made from 5 oz (150 g) of mixed butter and lard, 9 oz (250 g) plain flour, $\frac{1}{2}$ pint (300 ml) water, 2 teaspoons baking powder and a pinch of salt. The dumplings will rise; turn them over with a fork as they do so. Let the fruit simmer for not more than 20 minutes, and serve the blueberries hot with cream.

BLUEBERRY GINGERBREAD

Mix ½ lb (225 g) molasses with ½ pint (300 ml) sour cream. To this mixture, add ½ lb (225 g) flour, ½ teaspoon salt, 2 teaspoons bicarbonate of soda, and 1 teaspoon ground ginger. Dust 5–6 oz (150–175 g) selected blueberries with 1 oz (25 g) flour, and fold into the cream and molasses mixture, place in a long, greased baking tin to make a depth of about 1½ inches (4 cm), and place in the oven. Bake at low to moderate heat (300–350°F/150–180°C/mark 2–4) for 40–50 minutes or until well-risen and firm to the touch, remove, and turn out to cool.

Cloudberry *Rubus chamaemorus*, a bramble found in northern and arctic parts of America and Canada, in Scandinavia and sometimes in the Scottish Highlands and Welsh mountains. The fruit is sweet and juicy, with bunched amber-yellow and red berries, and a taste likened to honey or to apples – they are known in Canada as baked-apple berries. They feature in tarts and similar confections in Scandinavian cookery.

Dewberry The American name for blackberries that are on trailing stems rather than upright bush blackberries; used in cookery in the same way as blackberries.

Elderberry The fruit of the elder, *Sambucus niger*, ripe when it turns from red to deep purplish black, usually in September or October. The small berries are scrabbled off with a fork, a tedious job. The berries are traditionally used for syrup, vinegar, wine and elderberry sauce (the berries are baked with vinegar and then strained and cooked with a shallot, root ginger, cloves and peppercorns). The berries amalgamate well with apples to provide conserves or jellies. Germany has a soup *(Holundersuppe)* derived from the berries:

ELDERBERRY SOUP

Use ½ lb (225 g) ripe elderberries. Cook until soft in ½ pint (300 ml) water, and rub through a sieve to make a purée. Thin this by adding ¾ pint (450 ml) boiling water. Peel and slice ½ lb (225 g) apples, add them, and sugar to taste. Grate in the rind of a lemon, add a little salt, and cook until the apples are soft. Stir in the juice of the lemon and any more sugar that is needed. Before serving, skim off the lemon rind.

Hawthorn The dark-red berries, haws, are abundant in the autumn; a dry fruit, they make a good jelly, but need the addition of apples.

Huckleberry There are blue huckleberries *(Gaylussacia frondosa)* and red ones *(V. parvifolium)*, the red belonging to the north-west coast of America. The blue huckleberry, general in north America, is more prized for its aroma and flavour than many blueberries, and is used similarly for pies and for a huckleberry cake.

Rowanberry The mountain ash, *Sorbus aucuparia*, with brilliant orange berries is often cultivated as an ornamental tree. In Ireland, Wales, Scotland and Scandinavia the berries are used for jam and preserves – they are best allied with apples for this purpose. The American service berry may be regarded as the same as the sorb or rowan.

DONEGAL ROWANBERRY JELLY

Pick 1 lb (450 g) rowans, and gather 1 lb (450 g) crab-apples or tart cooking apples. Wash the apples, and slice them up without peeling them into a preserving pan. Add the rowanberries, and just cover the fruit with water. Boil gently until the fruit turns to pulp, then strain through a jelly bag, preferably overnight. Bring the juice back to the boil, first adding 1 lb (450 g) sugar for every pint (600 ml) juice and dissolving the sugar over a low heat, and continue cooking until the jelly will set (see jam section, p. 145). Put into hot jars and cover when cold.

Blackberry

THE blackberry is the most pleasant product of the wild fruit gallery. H. L. Mencken relates that his sole passion as a child was stewed blackberries, spread still warm on homemade bread, also still warm. The United States shares with Britain a love of the hedgerow bramble or blackberry, *Rubus fructicosus*, although the American bramble *(R. villosus)*, of which Whitman wrote that it would adorn the parlours of heaven, is a different variety. Both are cultivated as well as wild; there are many species. In Europe and Australia, the blackberry is also abundant. In France, however, *les ronces* were once entirely left to schoolchildren and peasants, and they hardly feature now except in jam or jelly.

An unusual feature of the blackberry bush is that blackberries are green, red and black on the stem at the same time. In September the lowest berry ripens alone, large and juicy, for eating raw; later in the month, secondary berries in the cluster ripen to a firm small size, while in October the last berries mature, but these are small and have predominant seeds. Blackberrying, with the attendant prickles, is thus particularly attractive, for the bush can be returned to twice for further fruit and to exercise the right to eat, while picking, the over-ripe and crumbling berries hidden deep in the bush. It must be one of the few fruits that draw town-dwellers into the country for a day among the hedgerows.

Many country virtues are imputed to the blackberry and its bush, as are a number of country superstitions. There used to be in Sussex, a county that made much blackberry wine, a taboo against eating blackberries after 10 October, the night on which the devil passes by and spits on every bush. The tenth of October is in fact 29 September in the Old Style calendar, and this is Michaelmas Day, when Lucifer was ejected from heaven. Blackberries are certainly almost over by then.

BLACKBERRY SOUP

Carefully wash 1½ lb (675 g) ripe blackberries, and set aside a few whole ones. Peel, core and slice up an apple. Put the fruit into a pan with 1 pint (600 ml) water, add half a stick of cinnamon and bring to the boil. Simmer until the fruit is tender, then rub through a sieve, add sugar as necessary, and thicken with a paste made from not more than 4 tablespoons potato flour mixed with a little water. Stir, bring to the boil again, and remove from the heat. Have a meringue mixture ready prepared from stiffly beaten egg white and 2 tablespoons icing sugar, and spoon this in portions on to the hot soup. Cover and leave, then chill thoroughly before serving, adding, finally, the few whole blackberries.

BLACKBERRY AND APPLE PIE

Peel, core and cut up 1 lb (450 g) cooking apples into quite small pieces. Wash 12 oz (350 g) blackberries, and mix the fruit together to fill a pie dish. Put an egg cup in the centre of the dish, sprinkle a good 4 oz (100 g) castor sugar over the fruit, and cover with short-crust pastry (see p. 11). Prick round the edges. Bake in a moderately hot oven (400°F/200°C/mark 6) for 15–20 minutes, and then at 350°F/180°C/mark 4 for a further 15–20 minutes.

BLACKBERRY BATTER PUDDING

To prepare a batter, mix together 3 oz (75 g) butter, 2 tablespoons sugar and 2 beaten eggs, then alternately add ½ pint (300 ml) milk and 9 oz (250 g) plain flour, sifted with 2 teaspoons baking powder and 2 pinches of salt, and beat well. Butter and lightly flour a round cake tin, and in it put a layer of the batter. Cover this with 10 oz (275 g) blackberries sprinkled with castor sugar. Pour the rest of the batter on top of the fruit. Bake in a moderate oven (375°F/190°C/mark 5) for half an hour, or until well risen. Arrange small dabs of butter over the top, sprinkle with brown sugar, and melt this under the grill. Serve with cream.

BLACKBERRY COBBLER

Place 1½ lb (675 g) washed blackberries, with a little sugar, in a buttered pie dish. Sprinkle the fruit with 1 tablespoon lemon juice, and dab with 1 oz (25 g) butter. Rub 2½ oz (65 g) lard into 9 oz (250 g) plain flour, and sift with 4 teaspoons baking powder and 1 teaspoon salt; stir in 1½ tablespoons castor sugar and 2½ tablespoons milk, or a little more if needed, and roll out the pastry to fit the pie dish. Cover the dish, trim away the edges, and prick with a fork. Bake for half an hour in a hot oven (425°F/220°C/mark 7), and serve with cream.

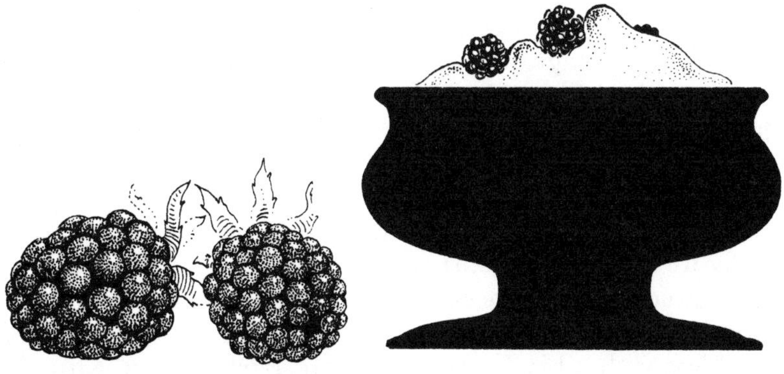

BLACKBERRY AMBER

Stew 1 lb (450 g) blackberries and sweeten them; put the fruit and some, but not all, of the juice in a buttered pie dish. Beat up an egg yolk with a little milk and in this dip enough slices of sponge cake to cover the fruit. Bake the dish in a moderate oven (350°F/180°C/mark 4) until the top browns. Then remove and cover with the white of an egg whipped to a froth with 2 oz (50 g) castor sugar. Turn down the oven to slow (275°F/135°C/mark 1), and replace the dish, leaving it until the cake is amber in colour. Serve hot, with cream.

NAMES: Fr. *mûre*; Ger. *Brombeere*; It. *mora di macchia*; Sp. *zarza mora*.

STORAGE: Blackberries spoil quickly, and once washed of the dust they seem especially to attract should be eaten raw, or cooked or prepared, the day they are gathered.

Blackcurrant

THE blackcurrant, *Ribes nigrum*, is a European bush fruit, not commonly found wild except perhaps in some damp north country wood. Distinctly fragrant in bud and leaf, a well-tended bush will produce ample strings of fruit from June to August. Prized as a fruit in Britain (a good blackcurrant pie is held to be rivalled only by the traditional gooseberry pie), the blackcurrant has received scant attention from growers in other countries. A wild variety is native to America, but is not cultivated; the French *cassis* is used almost exclusively for the drink, a sort of ratafia, of the same name, or for *Kir*, the agreeable mixture of white wine and blackcurrant juice. In Burgundy, a *confiture de cassis* is made, and there are *fondants de cassis*, sweets filled with a blackcurrant liqueur, *crème de cassis*, with which Dijon is associated. Further north, the town of Bar-le-Duc in particular lends its name to a blackcurrant (or redcurrant) conserve. Russia has jams, jellies and creams that use the fruit, a wine made from the berries and a drink made from the leaves – the region of the Baltic sea was one of the earliest areas of cultivation in Europe.

Yet it seems that the blackcurrant was not always popular in England; both Gerard and Parkinson speak of their dislike of the fruit, Gerard describing 'a stinking and somewhat loathing flavour'. Nevertheless, the blackcurrant has long provided country people with cordials and drinks from its juice that are abundant in vitamin

C. The synonyms 'quinsy berries' and 'squinancy berries' indicate the value of the fruit in treating sore throats. Regional cookery has many attractive tarts, jellies, jams, pies, fools and sorbets, and a blackcurrant sauce features in an old Cheshire recipe. One of the best ways of eating blackcurrants is raw with rum and brown sugar; the recipes for cooking them are straightforward ones.

BLACKCURRANT TART

Line a shallow pie dish with shortcrust pastry (see p. 11), then sprinkle with finely crumbled fresh breadcrumbs. Spread evenly over this ½ lb (225 g) ripe, washed blackcurrants, and sprinkle with a few drops of lemon juice and then with 2 oz (50 g) sugar. Bake in a moderately hot oven (400°F/200°C/mark 6) for 15 minutes. Remove, and pour over the fruit ½ pint (300 ml) milk, in which an egg has been beaten up and to which 2 oz (50 g) sugar and a touch of cinnamon have been added. Return the tart to the oven and continue baking at a moderate heat (350°F/180°C/mark 4) for 20 minutes. Sprinkle the top with a little castor sugar before serving with cream.

Here is an old-fashioned, boiled batter pudding, a recipe that can equally be used for damsons, redcurrants, green gooseberries or cherries:

BLACKCURRANT PUDDING

Prepare a batter from 4 tablespoons flour, 2 large eggs, and ½ pint (300 ml) milk and set to one side. Now butter a 2-pint (1·2-litre)

basin thickly, and fill it about two-thirds full with ripe blackcurrants. Pour the batter over the top. Butter and flour a cloth, tie it over the basin, and set the pudding in a saucepan of boiling water. Boil for about $1\frac{1}{2}$ hours, then turn the pudding out on to a hot dish, sprinkling plentifully with castor sugar.

BLACKCURRANT CHEESE

Place stalked ripe blackcurrants in equal measure with lump sugar in a pan, and stir together over a low heat until the sugar has dissolved. Then bring very slowly to the boil, continuing to stir the mixture. From time to time remove the scum as the currants simmer for 50–60 minutes. Stir often. They should be very soft and pulpy. Rub the fruit through a fine sieve, return the fruit purée to the pan and bring to the boil while stirring. The blackcurrant can then be removed, put into small warmed pots. Cover tightly, when cold, as with jam (see p. 145).

NAMES: Fr. *cassis* (the bush is known as *groseiller*, a name shared with the gooseberry and other currants); Ger. *Johannisbeere*.

STORAGE: Should be used as soon as possible after picking unless preserved in sugar.

Breadfruit

BOTANICALLY, the tree *Atrocaprus communis* is part of the mulberry family; the seedless, cultivated variety (it is nowhere found wild, and the form of its lack of seeds indicates cultivation from the earliest ages) is called the breadfruit, the seeded is usually called the breadnut. A cosmopolitan fruit, the breadfruit is believed to be a native of the Malay Archipelago, and was first encountered by Spanish voyagers to the Solomon Islands; in 1686 Dampier described the plant after observing it in Guam. Captain Cook's ad-

vocacy of its introduction to the West Indies resulted in the notorious voyage of the *Bounty*, under Lieutenant Bligh, which reached Tahiti in 1788. A thousand breadfruit plants were potted and tubbed on board, but they were never transplanted in the West Indies because of the mutiny that led some of the crew home via the East Indies and the mutineers to Pitcairn's Island. The expedition was repeated by the *Providence* in 1792, and trees were landed at St Vincent and Jamaica. In spite of this, the Jamaican slaves, for whom the breadfruit was intended as cheap nourishment, disliked the food and refused to eat it.

Apart from Tahiti, where its existence is ascribed by legend to a miraculous intervention to stave off famine (although it does not in fact hold the economic rank of the *fei*, or wild banana), the breadfruit grows well in Sri Lanka (Ceylon), Burma, Hawaii, India, the West Indies and Central America. It cannot be cultivated in the United States, although specimens of the tree are reported from Florida. The Fiji Islands, where it has been long established, distinguish eleven sorts of breadfruit tree; they are all handsome, lofty trees with large deep-cut leaves, and green or yellow fruits as large as Dutch cheeses, roughly tuberculated on the outside, weighing up to 10 lb (5 kg), and up to 8 inches (20 cm) in diameter. The Fijians use the word for 'heart', *uto*, to indicate the fruit; the tree itself has always been important in native cultures, the soft resinous wood being used for canoes, and the bark beaten into cloth. It has been calculated that one or two breadfruit trees will support one person for one year. Breadfruit can be boiled (rather like vegetable marrow), stuffed, fried in oil, or baked – baking in an oven produces the 'bread', although it is when it is unripe and white that the fruit is most nearly like bread; fully ripe, the pulp is yellowish. It is in fact closer to the potato, with its high starch content. The traveller A. R. Wallace recommended it as a pudding with sugar, milk, butter or treacle; fried in slices, it has been likened to Yorkshire pudding. Breadfruit can be obtained tinned, but must be fully drained before cooking.

CRÈME D'ARBRE À PAIN

Chop up 2 medium-sized onions into fine slivers, and chop up finely 1 clove of garlic, and heat these with 1 oz (25 g) butter in a frying pan until softened. Peel 1 fresh breadfruit, remove the core, and slice off ½ lb (225 g) of it. Dice this, and place with 1½ pints (900 ml) chicken stock in a saucepan. Add the onions and garlic, cover, simmer until tender, then use a blender to produce a smooth soup. Season with salt and pepper, stir in ½ pint (300 ml) cream, and reheat without boiling if the soup is to be served hot, or cool before blending and adding the cream, to make a breadfruit *crème Vichysoisse*. The latter is chilled thoroughly before serving and treated to a sprinkling of chopped chives – a pleasanter dish than a hot cream soup.

BREADFRUIT PUDDING

Peel and core a breadfruit, take 1 lb (450 g) of it and boil in salted water for 20 minutes or until tender. Drain and mash until smooth, and stir in 4 oz (100 g) sugar, 2 oz (50 g) melted butter, 2 tablespoons plain flour, and the grated rind of a lime. Mix well, also stirring in ½ teaspoon vanilla essence, ¾ pint (450 ml) milk, 2 eggs, and 3 tablespoons best rum. Beat into a smooth mixture that can be poured into a buttered dish then cook until set – about 45 minutes – in a moderate oven (350°F/180°C/mark 4). Serve with whipped cream, sweetened and flavoured with rum.

The following is an Australasian dish, part of the ceremony for welcoming visitors. Coconut cream, used in many island dishes, is made from grated coconut stirred in boiling water, and left to cool. A cream is strained out of this infusion into a bowl. When the bowl is refrigerated for half a day, a thicker cream rises to the top, and this is then skimmed off for use:

BREADFRUIT WITH COCONUT

Bake a whole breadfruit in the hot ashes of a fire, and let the skin turn black and crisp. The pulp will be soft. Peel the breadfruit as quickly as possible, cut the pulp into small pieces, coat it with coconut cream (known as 'lolo' in Fiji and its neighbouring islands) and serve. The skin, scraped or brushed clean, can also be eaten and has been compared in taste to the skin of a baked potato.

NAMES: Fr. *arbre à pain*; It. *albero del pane*; Ger. *Brotfrucht*; Dutch *broodvrucht*; Sp. *fruta de pan, pan de pobre*; Brazilian *fruta pão*; Solomon Islands *bulia*; Fiji *uto*; Hawaii, Samoa *'ulu*; Tahiti *maiore*.

STORAGE: It does not keep, and should be eaten when picked, preferably approaching full ripeness. Some Oceanic islanders store quantities buried underground for long periods, where the fruit assumes a cheese-like texture, and is used to make *madrai*, or native bread. Breadfruit does not travel well, although sound imported fruits can be bought in markets that cater for tropical tastes.

Cherry

THE cherry is a perfect-looking fruit, which can be yellow, streaked, carnelian red, or black, thin-skinned, luscious and bitter-sweet. It has long been a symbol of sensual desirability in our language; Herrick's blushing country girls were always cherry-ripe, and the fruiting orchards of the land were epitomes of fleeting pleasure. Gower says the old friars used to teach that 'life is but a cherye-fayre'. Real cherry fairs in provincial England must have been occasions for short-lived indulgence, much merry-making and consumption of local cherry pies and turnovers and cherry beer.

The fruit itself has a long history that probably started in Anatolia, the earliest home of cultivated cherries being the Mesopotamian orchards. In the eighth century BC, Sargon II of Assyria is said to have enjoyed the fragrance of their fruit. The Akkadian word *karshu* may be the source of the Greek *kerasos* and many similar names, but then the cherry may derive from the name of the Pontine city of Cerasus (the Turkish Gireson on the Black Sea), and if we are to believe Pliny no cherries were cultivated before 75 BC. The Romans are said to have introduced the tree to Britain: perhaps it died out, for it is recorded that Harrys, Henry VIII's fruiterer, did 'with no small labour bring from beyond the seas the sweet cherry, the temperate pippin, and the golden reinette' into Kent.

Whatever its origins, the cherry is part of the genus *Prunus*, the plum family, and is known by only two cultivated sorts in the temperate regions of the world: the sour and the sweet. All sour cherries, deriving from *Prunus cerasus*, belong to one of three groups: Amarelles, Morellos and Damasca. Amarelles are pale and bitter, but just edible raw – the Montmorency cherries belong to this group. Morellos, mainly divided into Black or Red Morellos, are acid and dark, and known to the French as *griottes*. The Dalmatian cherry is small, nearly black and with red flesh. There are many more varieties of sweet cherries, cultivated from *Prunus avium*, though they belong in general to one of three further groups: Geans, Bigarreaus, and Dukes. Geans or Guignes are either the Black, with dark skin and dark, sweet flesh, or the Red with a pale skin; Bigarreaus, also known as Hearts, are either of black or white skin, and Dukes (they originated in Medoc and are called *Royale* by the French) are a hybrid between sour and sweet cherries; again, these fall into further categories dictated by their colouring.

Although a large and richly fruiting tree, the cherry is fastidious about soil and climate – hence the fame of favoured areas in Kent and Buckinghamshire. It is also susceptible to canker and other diseases, and indeed its own excessive growth has drawbacks, one of which is that the best cherries are at the top of the tree, and handpicking is still the rule with cherries.

In their season, then, cherries of any kind are worth acquiring to eat as they are or to be made in the kitchen into one of many commendable dishes: cherries make good jams, tarts, pies, jellies, cakes, savarins and compotes, excellent sauces for game and poultry and a good accompaniment for meats. Not the least virtue of the cherry is that it is used to manufacture *kirschwasser*, the best sorts of which are from the Black Forest or Alsace. Kirsch has the merit of enhancing almost any fruit; a bottle of *kirsch de cuisine* is indispensable to the cook. From a cherry, too, a marasca distilled with honey, comes maraschino, a liqueur long associated with Zara in Dalmatia. English drinks once made from cherries include cherry-beer (in Kent) and cherry-bounce, a form of brandy.

CHERRY AND ROSE PETAL SALAD

Stone a number of ripe, sweet cherries and in place of the stones put cottage cheese; serve on lettuce with finely sliced tomatoes and spiced vinegar (see recipe for cherries in pickle, p. 77). Scatter rose petals over the salad as a garnish.

CHERRY AND HAZELNUT SALAD

Stone a quantity of large sweet cherries, and blanch an equal number of hazelnuts. Fill the cherries with these, and arrange on hearts of lettuce. Express some juice from a few extra cherries, and add to French dressing.

One of the pleasures of Hungarian food is a chilled soup of cherries. Here is one recipe for it:

CHERRY SOUP

Stone 1½ lb (675 g) cherries and place in an enamelled saucepan. Cover with 2 pints (1·2 litres) water and bring to the boil. Add 4 oz (100 g) sugar. Boil until the fruit is soft. Make a purée by rubbing through a sieve, return to the pan, and stir in 1 tablespoon arrowroot that has first been made into a smooth paste with cold water. When clear, remove from the fire, add the juice of half a lemon, and more sugar if necessary. Add 1 tablespoon cherry brandy. Serve cold, or – preferably – iced.

Ducklings are in season at the same time as cherries, in July and

August, and their delicate flesh is well set off by bitter fruits – oranges, apples, even fresh pineapple – and particularly sour cherries, of which the dark-fruited Morello is the best and the closest to the French cherry first cultivated to the north of Paris, at Montmorency.

CANETON MONTMORENCY

Prepare and roast the duckling – or 2 small ducklings – in the usual way, and set the pan juices, skimmed of fat, to one side. Melt in a pan 1 tablespoon butter, stirring in 1 tablespoon flour as it turns brown, then add $\frac{3}{4}$ pint (450 ml) prepared duck giblet stock and leave the mixture to simmer. Using another saucepan, stone 1 lb (450 g) bitter cherries, and add 2 tablespoons white sugar, a pinch each of ground cinnamon and nutmeg, the juice of an orange and its finely grated rind, and 8 tablespoons port; let this boil, then cover and simmer until the cherries are cooked. Remove the cherries, reduce the liquid until it begins to thicken, stir it into the stock, add the pan juices and the cherries, and heat the whole until ready. Serve the duckling with some cherries that have been reserved and a little sauce; serve the rest of the sauce separately.

Various parts of Britain have traditionally had different batters and batter puddings. In Kent, a basic batter, perhaps lightly flavoured with nutmeg, was cooked over stoned cherries and served, strewn with sugar, from the pudding basin; the version that follows is that associated with Tunbridge Wells. It is reminiscent of a number of French regional sweets, such as *Le milliard de cerises*, from the Auvergne, which is a pancake batter mixed with stoned black cherries and baked in a pie dish; *Le millia*, from the Nivernais, which consists of black cherries covered in a custard made of flour, water, eggs, salt and sugar, then covered with nuts of butter and baked and served hot; or a dish, not strictly a batter, from Franche-Comté, which consists of milk rolls soaked in milk and beaten eggs, mixed with stoned cherries and sugar, then moulded and baked. It is called *Roncin aux cerises de Montbeliard.*

KENTISH CHERRY BATTER

Boil 1 pint (600 ml) milk in a saucepan, and into it stir flour sufficient to make a thick batter. Beat, pour out, and, when nearly cold, beat in 3 eggs, a little sugar and flavouring. Allow the batter to cool in the dish, then cut into slices. Egg and breadcrumb each slice, using fresh breadcrumbs, and fry in lard. Serve on hot baked cherries.

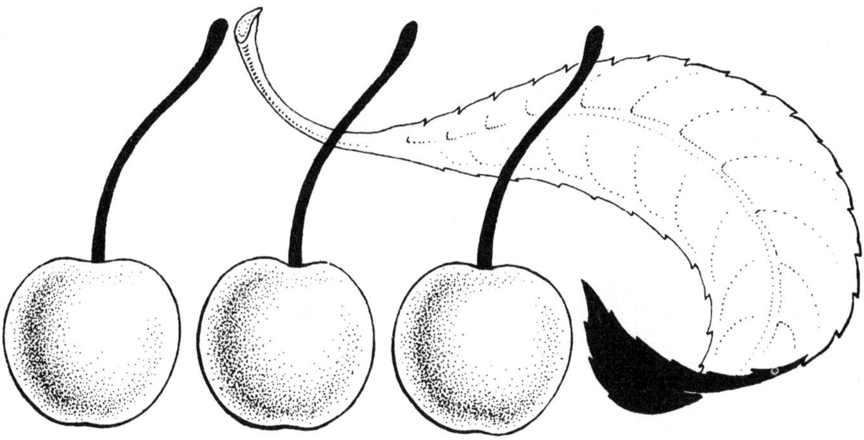

CHERRY PANCAKES

Make 8 pancakes from pancake batter (see recipe for Crêpes Sophie Victor, p. 45), and set them to one side. Make a syrup of $\frac{1}{2}$ pint (300 ml) water and 6 oz (175 g) sugar, and in this poach 1 lb (450 g) firm, red cherries, stoned, for 6 minutes. Drain off the syrup, and simmer to reduce until thick. Blanch 1 oz (25 g) almonds (see recipe for salted almonds, p. 15), shred them, and toast until brown. Now take a shallow ovenproof dish and place one of the pancakes in the bottom. Cover with a layer of cherries, then another pancake, and so on until the cherries are used up. End with a pancake on top, and pour over this the syrup, letting it settle round the layers. Bake in a hot oven (425°F/220°C/mark 7) for 15 minutes. Take out and serve while still hot, cutting the pancakes through like cake, and strewing almonds over each slice. Thick cream completes the helping.

Cherry

From Buckinghamshire, a county noted for its black cherry orchards and an annual Cherry Pie Feast at Cadsden, comes this recipe for cherry-filled pastries:

CHERRY BUMPERS

Take 12 oz (350 g) black cherries, stalked, washed, stoned and sprinkled with granulated sugar, and shortcrust pastry made with ½ lb (225 g) plain flour, ½ teaspoon baking powder, a pinch of salt, 4 oz (100 g) lard, and cold water to mix (see p. 11). Roll out the pastry to about ⅛ inch (¼ cm) thick, and cut into rounds of 4 inches (10 cm) across. Heap as many cherries as possible on to each round, damp the edges and fold over into turnovers, crimping the edges tight. Bake in a hot oven (425°F/220°C/mark 7) until the pastry turns golden brown. Sprinkle with castor sugar. Serve hot or cold.

In the following recipe, of 1671, 'ice it' means apply a sprinkling of sugar that will caramelize, 'a sirrup of Muskadine' means a syrup of either muscatel grapes or Malaga raisins, and 'Damask water' means rose water, properly from *rosa gallica var. damascena.* The cherries should be completely enclosed in very thin shortcrust pastry.

THE COUNTESS OF KENT'S TART OF CHERRIES

'Take out the stones, and lay the Cherries into your Tart, with Sugar, Ginger, and cinnamon; then close your Tart, bake it and ice it, then make a sirrup of Muskadine, and Damask water, and pour this into your Tart, strew on sugar, and serve it.'

CHERRY LAURETTE

First crush 2 lb (1 kg) raspberries in 1 pint (600 ml) cold water and strain them through a sieve. Make a syrup with 1 pint (600 ml) water and 1 lb (450 g) sugar. Let it cool, then add the raspberry juice and the juice of half a lemon. Freeze this mixture, folding in an egg white, beaten stiff, when it is half frozen. Meanwhile, melt 1 lb (450 g) redcurrant jelly with 2 tablespoons brandy, and add to this 2 lb (1 kg) large black cherries, stoned. Let these cool. When serving, pour some of the cherries over each portion of raspberry ice.

The following dessert was invented – and frequently consumed – by the great French actor, Coquelin, at the end of the nineteenth century:

COMPOTE DE CERISES COQUELIN

First take half a large pineapple, remove the outer skin and the hard centre and cut into thin slices; set these on a dish and sprinkle with sugar and kirsch, turning the slices from time to time to keep them fresh. Next stone 1½ lb (675 g) large ripe cherries. Put 12 fl oz (350 ml) Burgundy and ½ lb (225 g) sugar in a pan and let the sugar dissolve over a low heat, adding 20 cherry stones that have been crushed and enclosed in a muslin bag. Bring to the boil, add the cherries, cover, and simmer for 7 or 8 minutes until the cherries soften. Skim the cherries out into a dish and chill for half an hour. Meanwhile, reduce the syrup to just under half, strain it, let it cool, then stir in 1 tablespoon kirsch. Set the pineapple slices in a ring, and fill the centre with the cherries, adding the kirsch in which the rings have been soaking to the syrup. The dish can be served cold, or flambéed with kirsch. Serve the syrup separately.

Clafoutis, a regional dish of the Limousin and the Cantal, is authentically made as follows. The French leave the stones in the cherries of their *clafoutis*, and it is worth observing that where bulk and shape are important, as in puddings and pies, stones should be left in cooked cherries; doing so also preserves more flavour.

LE CLAFOUTIS

Take 3 eggs, 3 tablespoons flour, 3 tablespoons castor sugar, a pinch of salt and 1¼ pints (750 ml) milk. Put the flour in a terrine, then the salt, and one by one break the eggs into the flour. When you have stirred it, make a paste using a spoon, and pour in the milk, a little at a time at first, then the sugar. Beat well after each addition until a smooth batter is obtained. Next take an ovenproof dish and place in it about 1 lb (450 g) cherries. The darker they are, the better they will be (the Bigarreau is not a good cherry for a *clafoutis*). Pour over them the flour, egg and milk mixture, and place in a fairly hot oven (375°F/195°C/mark 5). Cook about 40 minutes or until set. When removed, sprinkle with sugar.

NAMES: Gk. *kerasus*; Latin *cerasus*; Old Norman Fr. *cherise*; Old Eng. *ciris*, *cyrs*; Fr. *cerise*; Ger. *Kirsch*; Dutch *kersen*, *krieken*; Arabian *garasiah*; Polish *trzesnia*; Russian *chereshnia*, *vishnya*, etc. Morello derives from *amarellum*, Med. Latin = small, bitter one.

STORAGE: In a cool, dry place, but for a short period only.

Chestnut

THE European, Sweet, or Spanish chestnut, *Castanea sativa*, is the impressive tree that yields the edible fruit and a useful, but not enduring, timber. There are two other chestnut trees. One, *Castanospermum australe*, is confined to Australia, where the aborigines soak, dry and roast the seeds, which are harmful raw: this is the Moreton Bay chestnut. The other is the widespread horse chestnut, *Aesculus hippocastanum*, probably originating in Greece, of which the fruit is inedible, although a substitute for coffee has been made from it, and animal fodder from its meal.

The Sweet chestnut has glossy, brown nuts, some up to 2 inches (5 cm) wide, protected by a long-spined green cupule; this is easily opened, but the shelling of chestnuts requires more attention. It is best to make an incision in the skin with a short-bladed knife and then to heat the chestnuts in the oven, or to fry or boil them for about 10 minutes, after which the outer and inner skins can readily be removed.

Widespread growth of the European chestnut, particularly in Italy, Spain, France and England, has made its fruit an important constituent of the diet of the mass of the people. When John Evelyn complained that in England it was mostly given to swine and the nut was really 'a lusty and masculine food for rusticks', he was probably aware of the extent of its use in Mediterranean countries. In England, the chestnut features little in regional dishes (a Cambridgeshire chestnut jam has been noted; see p. 147); the zenith of its consumption must have been in the mid-nineteenth century, when Mayhew counted more than a thousand roast chestnut sellers on the streets of London alone, and when the roasting of chestnuts on winter fire bars, or over hearths in those antique roasters that consisted of a long-handled perforated metal box that confined the exploding nut, was a popular pastime. Elsewhere, the chestnut has been exploited to the full for some 2,000 years, its value lying in its being highly farinaceous, with little oil, some vitamin B and C, and the desirable fat-soluble vitamin K. Chestnut flour, or a readily prepared *farine de châtaignes* that can be bought, means breads and puddings, cakes and purées; boiled, roasted or baked nuts form a vegetable-like accompaniment to meats. The versatile and substantial chestnut provides both the most basic dishes and the most elegant of *entremets* (see, among ice puddings, Bombe Nesselrode (p. 142) and Mont Blanc (p. 142), chestnut sauce (p. 244) and compote of oranges and chestnuts (p. 121)). The fruit deserves wider acquaintance than in the form of roast nuts or occasional *marrons glacés*.

CHESTNUT SOUP

Make an incision in the skin of 50 chestnuts and bake for 10 minutes. Shell and skin them, and place in a saucepan with 2 oz (50 g) butter; add a chopped onion, 4 lumps of sugar and a pinch of pepper and salt. Simmer them for 45 minutes, then remove the chestnuts and, keeping 6 whole ones on one side, mash them through a sieve. Return the pulp to the fire in another saucepan and add 2 pints (1·2 litres) brown stock and the 6 whole chestnuts. Serve very hot with *croûtons*.

Regarded here as a vegetable to accompany game or red meat, *castagne stufate* is one of many Italian chestnut preparations:

STEWED CHESTNUTS

Shell and skin up to 1½ lb (675 g) chestnuts, and cover them in a heavy saucepan with 8–10 fl oz (250–300 ml) of a dry white wine, such as Orvieto *secco*. Chop a medium-sized onion into this. Add 2 teaspoons of meat extract and a sprinkling of pepper, and top up with ½ pint (300 ml) water. Simmer gently for 45 minutes after covering the saucepan. Drain off the chestnuts, which should still be whole, into a warm dish. Sprinkle chopped walnuts over them, and serve immediately.

CHESTNUTS AND ONIONS

Shell and skin 3 lb ($1\frac{1}{2}$ kg) chestnuts, and boil until tender in a light vegetable stock made of potatoes, carrots, onions, leeks, celery, cabbage and a sprig of cauliflower. When tender, drain them on a sieve. Heat 3 oz (75 g) cooking fat and fry in it 3 chopped onions. Then add the boiled chestnuts, turning the mixture over gently as it cooks for a few minutes. Keep the chestnuts whole. Season with a pinch of salt and some nutmeg. Serve with onion rings fried crisply in butter.

The next recipe is a fruit accompaniment to be served with duck or goose, as in Germany. It can be further refined and served as a sauce by putting it through a strainer:

CHESTNUTS AND APPLES

Boil 1 lb (450 g) chestnuts, shell and skin them, then stew them in butter until nearly tender. Take 1 lb (450 g) apples, peel and slice them, and stew in $\frac{1}{2}$ pint (300 ml) white wine adding extra water if necessary until half done. Add 1 oz (25 g) sugar and the chestnuts, and cook together until well done. Serve.

PURÉE OF CHESTNUTS

Slit, bake, shell and skin 50 chestnuts. Place in a saucepan with enough white stock to cover. Boil slowly until the chestnuts are completely soft. Drain off liquid, then pass through a sieve. Add $\frac{1}{4}$ pint (150 ml) cream, a lump of butter, and $\frac{1}{4}$ pint (150 ml) of the stock in which they have been boiled. Season with salt and pepper, heat and serve.

A chestnut purée is the base of this poached egg dish:

POACHED EGGS À LA CEVENOLE

Poach 3 eggs in the usual manner, dry on a cloth and keep hot. Butter a dish, cover the bottom with a chestnut purée, and place the eggs on this. Cover them with a *sauce Mornay* – Gruyère, Parmesan and a light chicken stock, with a roux base – add a little more grated cheese and some butter and place the dish in a very hot oven (475°F/250°C/mark 9). Remove as soon as the top is golden brown.

BRAISED CHESTNUTS À LA MADÈRE

Shell and skin 1 lb (450 g) slightly roasted chestnuts, then fry them in butter. When brown, add ½ pint (300 ml) stock, pepper and salt, and 4 tablespoons good Madeira. Simmer until the chestnuts are tender. Pass the whole mixture through a sieve and heat again; before removing from the flame stir a little cream into the purée.

CHESTNUT PUDDING

Boil 1 lb (450 g) shelled and skinned chestnuts (about 2 lb (1 kg) unshelled weight) for about half an hour, or until they are soft, in water to which 1 tablespoon sugar has been added. Drain the chestnuts and pass through a sieve, then mix in 2 oz (50 g) butter and 2 oz (50 g) sugar. Stir this mixture over the heat for 5 minutes, then allow it to cool. Beat the yolks of 2 eggs and add these, then add the stiffly whisked whites of the eggs. Put the mixture into a greased mould, cover with buttered greaseproof paper or foil, and steam for 1½ hours. Turn out, and serve with a fruit sauce, or with stewed apricots.

And here is the Spanish version:

FLAN DE CASTAÑAS

Take 1 lb (450 g) chestnuts and make an incision in each before boiling for 10 minutes. Then shell and skin them, cover, and boil in milk with 1 tablespoon sugar, a pinch of salt and a little vanilla essence. When soft, mash and sieve the chestnuts. Stir together thoroughly the yolks of 4 eggs and 3 tablespoons sugar, and add to the chestnuts. Beat the whites of the eggs until stiff, and fold these in. Pour the purée mixture into a pie dish lined with caramel (see recipe for banana tart, p. 43), and stand this in a deep pan of hot water which is kept simmering. When a knife comes out clean, the pudding is done. Turn out so that the caramel is on top. Serve cold.

RICH CHESTNUT PUDDING

Pound together 3 tablespoons crushed *marrons glacés* until they form a paste, then pass them through a sieve into a bowl. Stir in ½ oz (15 g) butter, and stir briskly for 5 minutes. Add 1 oz (25 g) castor sugar, and 2 whole eggs, and stir for 5 minutes more until the mixture is creamy. Next add 2 oz (50 g) raw beef marrow, ½ oz (15 g) cake crumbs, 1 tablespoon cornflour, 2 tablespoons currants, and 1 full tablespoon kirsch. Stir rapidly for 1 minute. Lightly butter 6 individual pudding moulds and fill with the mixture; place them in a baking tin filled with hot water to half the height of the moulds. Put this in a hot oven (425°F/220°C/mark 7), and bake for 20 minutes. Unmould the puddings on to a warmed dish, and serve with a hot pineapple sauce, made by boiling diced pineapple in sweetened water until it is syrupy, then adding 1 tablespoon kirsch.

CHESTNUT AND CHERRY TART

Using good shortcrust pastry (see p. 11), make a flan case and bake blind. When cool, fill the bottom with sour red cherries, stoned. Morellos are suitable for this. Next prepare a smooth chestnut purée (see p. 68) and spread it evenly over the whole tart. Sprinkle castor sugar over the surface quite thickly, and grill the tart until the sugar caramelizes. Serve cold.

CASTAGNE AL MARSALA

Make the usual cut in the skins of 1 lb (450 g) chestnuts and place in a baking tin. Roast in a moderate oven (350°F / 180°C / mark 4) for 15 minutes, then remove both the shell and the inner skin, leaving the chestnuts in one piece. Arrange in a pan and cover with marsala. Sprinkle 2 or 3 tablespoons sugar over the chestnuts. Simmer very slowly until they are tender, and the wine has reduced to a syrup. Allow the chestnuts to cool, and serve with cream.

Finally, a Brazilian sweet of boiled chestnuts:

DOCE DE CASTANHAS

Shell and skin a quantity of chestnuts, and place as many as possible in a pan in a syrup made from 1 lb (450 g) sugar and 1 pint (600 ml) water. Boil the chestnuts until they soften, drain and sieve, then beat in the yolks of 2 eggs. The chestnuts should form a fudge-like mass when tested in cold water. When ready, pour the mixture into a greased baking tin and leave to cool. This is served as a sweet, sprinkled with cinnamon and with cheese to accompany it.

NAMES: Fr. *marron*; Ger. *Kastanie*; It. *castagna, morrone*; Sp. *castaña*.
STORAGE: Unshelled, chestnuts will keep for some weeks.

Chinese Gooseberry

THE Maori name for the flightless bird *Apteryx mantelli* is kiwi, and some fancied resemblance has given the alternative name of kiwi-fruit to what is properly the Chinese gooseberry. The fruit is from a sub-tropical vine, *Actinidia sinensia*, originating, as its name indicates, in China, and principally cultivated in California and New Zealand. The edible fruit is cylindrical and lemon-sized, brown or greenish-brown and fuzzy; the inside has a slightly transparent melon-like flesh, dark green radiating from a pale centre, with small black seeds. A variety known in Japan, *Actinidia callosa*, is only 1 inch ($2\frac{1}{2}$ cm) in length, and has a very delicate flavour; the Japanese simply name it *kōkwa*, 'good fruit'. The Chinese gooseberry is indeed a very good and tempting fruit, and it has considerable merit as a food, with a superabundance of vitamin C, and elements of protein, calcium, phosphorous salts and iron.

The skin must be rubbed to rid it of the fine hairs, and it is best peeled if it is to be sliced. If it is to be left unpeeled, halve the fruit, sprinkle it with lime or lemon juice, and eat with a spoon. Chinese gooseberries make a sound addition, sliced, to fruit salads. In New Zealand, an 'Auckland steak' is prepared by first rubbing fillet or

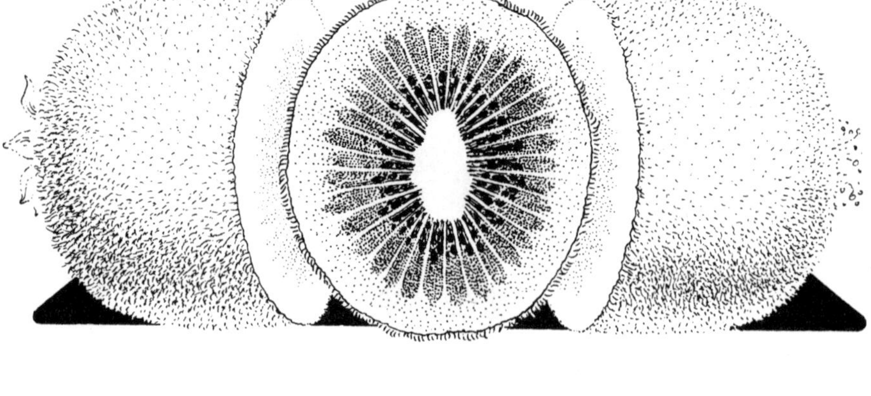

sirloin beef with a peeled kiwi fruit to impregnate it with juice; the grilled meat is then served garnished with the fruit in slices.

CHINESE GOOSEBERRY FRUIT SALAD

Place the flesh of six Chinese gooseberries in a basin with the diced flesh of 1 ripe papaya. Add $\frac{1}{4}$ pint (150 ml) sweet sherry, and leave to chill for 1 hour. Cut in half 3 small ripe cantaloupes, scoop out the seeds, and fill with the fruit mixture. Serve with castor sugar.

Kiwi-fruit Pavlova is a fruit meringue justly popular in Australia and New Zealand, and said to have been contrived to celebrate the ballerina's tour of those countries, in 1925. The filling is more often of passionfruit pulp folded into whipped cream; strawberries or other similar berries will also serve.

CHINESE GOOSEBERRY PAVLOVA

Make the meringue case by first beating the white of 4 eggs until stiff enough not to move, then adding 8 oz (225 g) castor sugar, a tablespoon at a time, and beating well until thick. Gradually add $\frac{1}{2}$ teaspoon white vinegar, then sift 1 teaspoon cornflour over the mixture and fold in thoroughly. Turn the mixture on to a greaseproof paper-lined tray, form into a round and hollow out the middle. Bake for about $1\frac{1}{2}$ hours in a very slow oven (275°F/135°C/ mark 1). When crisp, and before it colours, remove the Pavlova from the oven. It will be brittle while still hot, and care in handling and removing the paper is essential. When quite cold and ready to serve, whip $\frac{1}{2}$ pint (300 ml) double cream until thick, fold in the pulp of 1 skinned Chinese gooseberry, and fill the meringue. Decorate with finely sliced rings of at least 2 more Chinese gooseberries.

CHINESE GOOSEBERRY PIE

Line a pie dish with shortcrust pastry (see p. 11), and bake for 10–15 minutes in a moderately hot oven (400°F/200°C/mark 6), then leave it on one side to cool. Prepare a filling of about 8 peeled and sliced Chinese gooseberries; sprinkle them with 4 teaspoons lime juice. Mix 4 oz (100 g) sugar with a pinch of salt and some grated lime peel. Toss the fruit slices in this mixture and turn them into the pastry shell. Heap them so that the centre is raised, and sprinkle over them a topping composed of blended 3 oz (75 g) flour, 1 oz (25 g) sugar and a little salt, with pea-sized pieces of butter (about 1 oz (25 g)) cut into it. Bake in a moderate oven (350°F/180°C/mark 4) for 10 minutes. Reduce the heat and continue cooking until the crumb topping is brown. Serve hot with cream.

STORAGE: The Chinese gooseberry should be bought when firm and allowed to ripen at room temperature until it yields to a slight pressure. It will then keep for about a month refrigerated.

Chutney, Ketchup and Pickles

PRESERVING perishable foods of all kinds, at best from season to season, and at least from day to day, is as ancient a preoccupation as cookery itself. All the older cookery and receipt books are replete with evidence of a struggle against the processes of time, a battle that is no longer necessary in an age of refrigeration. But the tradition of 'preserves', of reserving food for future occasions, has been so strong in Western cuisines that it has imposed patterns of eating; to a lesser degree, this is reflected in chutney, ketchups and pickles, all of which rely on fruits or vegetables. Pickles have always been widely known; the Romans, who made a preserving substance called *garum* or *liquamen*, supposed to be of fermented fish, possibly the gills and blood of a tunny, and the Chinese, who used a brine of pickled fish which in Amoy dialect was called *ké-tsiap* (hence our ketchup), were forerunners of our simpler usage of vinegar, sugar and spice. The Indian chutney, probably now bearing little resemblance to the Hindi *chatni* brought back by the Army or East India Company in the late eighteenth century, is a hot blend of fruits and condiments; pickles are fundamentally whole fruits or vegetables preserved, and ketchups are preserves verging on sauces. Apart from the virtue of keeping for long periods if properly prepared, all these are excellent accompaniments to curries, roast meats, cold meats, salads, or bread and cheese.

Some caution is required in making these preserves. Decay being the natural latter-end of ripeness, it is imperceptible in its beginnings, and for this reason 'ripe' fruits should be those that are barely so, and not quite ripe fruits are to be preferred. Any suspicion of doubt about damaged or bruised fruit, or even over-soft fruit, must be assuaged by rejecting it as unworthy of preserving. Equally, bottles and jars must be clean and sterile, and the sealing of them air-tight and flawless. Vinegar-proof lids must be used. No uncoated metal should be allowed to touch the preserve. Paper covers are useless because they will allow the mixture to evaporate. Spices should be ground, or at least bruised. Sugar for dark chutneys should be brown, otherwise either granulated or brown may be used. Vinegar means fundamentally malt vinegar, but wine vinegar provides an enhanced flavour, as do vinegars spiced by infusing herbs such as tarragon, thyme, or marjoram for at least two weeks before use. Pans should not be made of a metal affected by vinegar (aluminium, stainless steel or enamel are suitable), and wooden spoons and spatulas must be used. A last point to remember about all preserves is the accurate labelling of the jars with the description and the date made.

APPLE AND ONION PICKLE

In $\frac{1}{2}$ pint (300 ml) vinegar, steep 12 chillies, 6 cloves and 6 peppercorns, and add 1 tablespoon salt. Simmer this for 10 minutes, strain, and cool. Now chop up 2 lb (1 kg) sharp apples and 2 lb (1 kg) onions into fine pieces. Pack into warm jars, and cover with the vinegar; the pickle is ready for use immediately.

APRICOT, APPLE AND PEACH CHUTNEY

Halve and stone 2 lb (1 kg) apricots, peel and core 2 lb (1 kg) green cooking apples, and blanch, peel, halve and stone 4 firm peaches. Chop all the fruit into pieces of equal size. Finely chop 2 onions and 2 oz (50 g) peeled root ginger. Combine all the ingredients in a large saucepan, together with the grated rind of 1 lemon and 2 oranges, the juice of the oranges, 1 teaspoon dry mustard, ¾ teaspoon grated nutmeg, ¾ teaspoon ground allspice and ½ lb (225 g) raisins. Add 1 pint (600 ml) wine vinegar. Bring the mixture to the boil fairly quickly, then simmer for up to 1½ hours, or until the fruit is soft. Stir in 1 lb (450 g) white sugar and 1 lb (450 g) brown sugar, and a further ½ pint (300 ml) vinegar, dissolve the sugar over a low heat, then continue to simmer for about 45 minutes, or until the chutney thickens appreciably. Ladle into warm jars and seal when cold; leave for 6–8 weeks before using.

APRICOT CHUTNEY

Peel 4 oz (100 g) root ginger and pound it to a smooth paste. Boil together 4 lb (2 kg) fresh stoned apricots and 2 lb (1 kg) sugar until thick, and add the ginger paste, together with 4 oz (100 g) chopped chillies, 2 cloves of garlic, chopped, 4 oz (100 g) salt, 2 lb (1 kg) stoned raisins, chopped fine, and 2 pints (1·2 litres) vinegar. Return to the boil. Cook for 10 minutes. Pack into warm jars and seal when cold.

BANANA CHUTNEY

Peel, core and slice 1 lb (450 g) tart cooking apples, add a little water, and cook to a pulp. Peel and chop up 1 lb (450 g) onions. Slice 4 oz (100 g) of crystallized ginger into fine pieces. Peel and slice 2 lb (1 kg) ripe bananas. Mix the cooked apple with the onions, ginger and bananas, then add 6 oz (175 g) brown sugar, 1 tablespoon salt, 1 teaspoon cinnamon, ¼ teaspoon cayenne pepper and ½ pint (300 ml) white vinegar. Simmer gently for about 1½ hours. Stir occasionally as the mixture thickens. Pack into warm jars and seal when cold.

BLACKCURRANT CHUTNEY

Pick from their stalks 1 lb (450 g) blackcurrants and cover with vinegar. Cook gently until they are tender, and let them cool. Then mix in the ingredients that make the chutney: 4 oz (100 g) chopped, seedless raisins, 1 oz (25 g) crushed mustard seed, 1 oz (25 g) chopped onion, $\frac{1}{2}$ oz (15 g) ground ginger, 2 oz (50 g) salt and 1 lb (450 g) moist brown sugar. Stir vigorously together. Return to the boil and cook for 10 minutes, then bottle, covering tightly when the chutney is quite cold.

CHERRIES IN PICKLE

Make a spiced vinegar by tying in a muslin bag 12 cloves, a 2-inch (5-cm) stick of cinnamon broken into pieces and a small piece of root ginger, immersing it in $1\frac{1}{4}$ pints (750 ml) vinegar and boiling in a pan with the lid on for 20 minutes. Meanwhile, fill some warm preserving jars with washed and dried cherries, leaving a short, 1-inch ($2\frac{1}{2}$-cm) stem on each, and arranging them in layers with a thick bed of icing sugar between each. The jars should only be three-quarters full; fill them up with the spiced vinegar, which has been left to get cold. Seal the jars down.

CHINESE GOOSEBERRY CHUTNEY

Wash 12 Chinese gooseberries, peel them and put them in a preserving pan. Peel 3 medium-sized onions and chop finely. Slice up 1 lemon, disposing of the seeds. Chop up 4 oz (100 g) seedless raisins. These ingredients should then be combined with the following: ¾ pint (450 ml) cider vinegar, 12 oz (350 g) brown sugar, 5 tablespoons of coarse-cut chopped preserved ginger, 1 teaspoon ground ginger, 1 teaspoon salt and ¼ teaspoon cayenne pepper. Place them all in the preserving pan, and simmer slowly and gently over a low heat for 1½ hours, stirring continually until a smooth pulp is obtained. Transfer it to warm jars and seal at once.

CRANBERRY KETCHUP

Wash 1 lb (450 g) cranberries, and simmer in an enamel saucepan with ¼ pint (150 ml) water and a chopped onion. After about 10 minutes make a purée of the cranberries, pressing them through a sieve with a wooden spoon. Return the purée to the saucepan. Stir in ¼ pint (150 ml) vinegar, ½ lb (225 g) sugar, and ¾ teaspoon each of ground cloves, cinnamon, allspice, salt and celery seed, and ½ teaspoon ground black pepper. Bring to the boil and cook for 15 minutes, stirring continuously and skimming off any surface scum. The ketchup should be thick and just drop from the spoon. Ladle it into warm jars, and seal when cold.

DAMSON PICKLE

Cook unstoned damsons, allowing 2 lb (1 kg) sugar to every 2 lb (1 kg) damsons, in a covered pan until the fruit is tender but unbroken. Strain off the syrup and add to it ½ pint (300 ml) vinegar, 10 peppercorns, 6 cloves, 6 berries of allspice and a pinch each of ground ginger and mace. Bring to the boil, and simmer for 10–15 minutes. When it has cooled, pour over the damsons, which have been placed in warm jars. Cover the jars and seal tightly, leaving them to mellow for 10–12 weeks.

DATE CHUTNEY

Place ½ lb (225 g) fresh stoned dates, and 1 lb (450 g) ripe, but firm, red tomatoes of small, uniform size in a saucepan, and to them add 4 oz (100 g) each seedless raisins and currants. Cover with ½ pint (300 ml) vinegar, add 1 teaspoon salt, and bring steadily to the boil, stirring the mixture constantly. Simmer the chutney for about 1 hour, stirring occasionally, until it thickens. Ladle into warm jars, and cover and seal when cold.

ELDERBERRY CHUTNEY

Wash 2 lb (1 kg) elderberries and remove from the stalks. Put them in a pan and bruise with a wooden spoon. Add 1 pint (600 ml) vinegar, a large onion, finely chopped, 1 teaspoon each of ground ginger, salt and mustard seed, 5–6 oz (150–175 g) sugar, and a pinch each of cayenne and mixed spice. Bring all this to the boil, and then let simmer until the chutney thickens. Stir thoroughly, put into warm jars, and seal when cold.

FIG PICKLE

First dissolve ½ tablespoon bicarbonate of soda in 4 pints (2·4 litres) water, bring to the boil and pour over the figs – 3½ lb (1¾ kg) will suffice, and the figs should be not quite ripe. Stand for 10 minutes, drain off the water, and rinse in cold water. Then dissolve 1½ lb (675 g) sugar in 2 pints (1·2 litres) vinegar, and add 2 small sliced lemons, 2 teaspoons cloves, ½ tablespoon ground cinnamon, and 1 tablespoon preserved ginger, chopped fine. Now add the figs. Boil until they are tender and clear, but do not let them split. Lift out the figs, drain thoroughly and put into warm jars. Simmer the remaining syrup until it thickens, and pour over the figs until they are well covered. Seal while hot. Keep at least a week before serving.

GOOSEBERRY CHUTNEY

Wash, and top and tail, 4 lb (2 kg) ripe green gooseberries, and chop them, together with 3 medium-sized onions. Place in a preserving pan and add the following ingredients: $1\frac{1}{2}$ lb (675 g) brown sugar, $\frac{3}{4}$ pint (450 ml) each cider vinegar and white wine, 6 oz (175 g) seedless raisins, and 2 teaspoons dry mustard. Add also 1 teaspoon each of salt, ground ginger and turmeric, and $\frac{1}{2}$ teaspoon cayenne pepper. Cook this mixture slowly, stirring it constantly, until it thickens, after about 2 hours. Ladle into warm jars and seal when cold.

GREEN FIG CHUTNEY

Cut up 3 lb ($1\frac{1}{2}$ kg) green figs into round slices. Dice 4 oz (100 g) preserved ginger and 6 oz (175 g) dates. Combine 2 pints (1·2 litres) vinegar with 12 oz (350 g) brown sugar, dissolve sugar, bring to the boil, then pour mixture over the figs, ginger and dates. Now add 6 oz (175 g) raisins, 1 lb (450 g) onions, chopped, 1 teaspoon salt and a large pinch of cayenne pepper. Leave the ingredients, when mixed, overnight, then boil for 3 hours. The chutney will thicken and turn dark.

LIME PICKLE

Take 25 fresh limes, and cut them into thin slices, removing the pips. Skin 6 oz (175 g) root ginger, and slice this equally finely. Chop up 2 oz (50 g) garlic. Pound 4 oz (100 g) dried chillies. Mix the spices with $\frac{1}{2}$ pint (300 ml) vinegar, season with both salt and sugar and bring to the boil. Cool, add the limes and bottle.

MANGO PICKLE

Take 2 lb (1 kg) fresh green mangoes and cover them with salt.

After a day, strain off the water that has exuded from the fruit, wipe the mangoes, and cut them half open to remove the seeds. Rub them with a little pounded turmeric mixed in 4 tablespoons salad oil. Next, prepare a stuffing by pounding together a mixture of 12 oz (350 g) green chillies, ½ lb (225 g) garlic cloves, ½ lb (225 g) root ginger and 10 oz (275 g) mustard seed; stuff the mangoes with this and bind the mango halves together with thread. Put the mangoes in a wide-mouthed jar and cover with vinegar which has been boiled and allowed to cool. Seal the jar.

NECTARINE CHUTNEY

Blanch, skin and stone 3 lb (1½ kg) ripe nectarines. Slice them, and place in a preserving pan. Add 1 finely chopped onion, 6 oz (175 g) seedless raisins, 5 tablespoons finely sliced preserved ginger, the juice of 2 lemons and ¼ pint (150 ml) cider vinegar. Put in 1 teaspoon ground allspice, ½ teaspoon each of ground cinnamon, ground cloves and ground ginger, and 1 tablespoon salt. Add 5 tablespoons pectin to ensure setting. Mix all these ingredients thoroughly, bring to the boil, and boil for 15 minutes. Add 1½ lb (675 g) white sugar and 12 oz (350 g) brown sugar. Dissolve the sugar, cook for another 5 minutes, then ladle into warm jars and seal when cold.

ORANGE CHUTNEY

Pare 3 oranges, chop the peel, cover it with water and cook until tender. Drain and keep on one side. Peel, core and dice 6 good-sized apples. Into a saucepan put the pulp of the oranges, cut into pieces, 4 oz (100 g) chopped citron peel, 4 oz (100 g) sultanas, 4 medium-sized onions cut into thin rings, the drained pulp of a 14 oz (397 g) tin of tomatoes, 1 lb (450 g) Demerara sugar, 2–3 oz (50–75 g) salt, 1 tablespoon each of black pepper and ground ginger, 2 teaspoons ground allspice, 1 teaspoon chilli powder, 4 oz (100 g) finely chopped crystallized ginger, and 1½ pints (900 ml) malt vinegar. Bring to the boil. Now add the orange peel and cook slowly for 1 hour. Add the apples, simmer for half an hour, and stir well. Set aside for 24 hours and then fill as many small jars as necessary and cover them. Store in a cool place. The chutney will be ready in a week.

PEAR CHUTNEY

Peel and core 3 lb ($1\frac{1}{2}$ kg) ripe, but firm, pears, and chop into manageable pieces of equal size. Do the same with 2 lb (1 kg) apples, not sweet ones, and put the fruit into a saucepan. Add 2 chopped onions and 1 lb (450 g) seedless raisins. Mix in a chopped chilli, a piece of chopped root ginger, a crushed clove of garlic, 1 teaspoon salt and $\frac{1}{2}$ teaspoon grated nutmeg. Add 6 cloves, and the juice and finely grated rind of 2 oranges. Put in 1 pint (600 ml) vinegar and 1 lb (450 g) moist brown sugar. Dissolve the sugar, bring the mixture to the boil, stirring frequently, then reduce the heat and simmer for 3 hours. An occasional stir is needed. When thick, ladle the chutney into warm jars and seal when cold.

PEARS, SPICED AND PICKLED

Peel 6 lb (3 kg) pears and dig out the core ends. Boil the fruit with 4 pints (2·4 litres) water until soft. Remove the pears. Add to the juice in the pan 2 pints (1·2 litres) vinegar, 1 tablespoon ground cinnamon, a piece of root ginger, 2 tablespoons ground cloves and 1 lb (450 g) sugar. Boil for 10 minutes, replace the pears, and simmer until the liquid thickens. Place the pears in jars; cook the syrup another 5 minutes and pour boiling over the pears. Cover and tie down when cold. More spices can be used if desired, and more sugar to give a thicker syrup.

PLUM KETCHUP

Stone 3 lb ($1\frac{1}{2}$ kg) plums, which should be firm and ripe, and preferably red and astringent, and place them in a saucepan with $\frac{1}{2}$ pint (300 ml) cider vinegar and $1\frac{1}{2}$ lb (675 g) sugar. Stir with a wooden spoon until the sugar nearly disappears, bring to the boil, then lower the heat to simmering until the plums can easily be squashed in the pan. Stir in 2 teaspoons ground cinnamon, $\frac{1}{2}$ teaspoon each of ground mace and nutmeg, $\frac{3}{4}$ teaspoon ground cloves and a pinch of ground black pepper. Boil for about 45 minutes, stirring occasionally to prevent sticking, skim, ladle into warm jars, and seal when cold.

WATER MELON PICKLE

Pare off the green part of the water melon rind, and cut the white part of it into strips of a size for pickling. Pour boiling water over these to cover, and leave overnight. Then soak for a few hours in cold water before placing in a preserving pan and cooking in water until tender. Meanwhile, prepare a syrup of equal parts of sugar and vinegar, and in this cook a few sliced lemons from which the seeds have been removed, a piece of root ginger and a stick of cinnamon. When the syrup thickens, drain the melon, place in the syrup, and cook until the melon is clear. Remove the melon and place in jars. Boil down the remaining syrup until it thickens further, then pour it over the melon rind and seal when cold.

Crab-apple

CRAB-APPLES, those cultivated and those growing wild in the hedgerows of England, on the heaths and in the woods of northern Europe and in America, are virtually all hybrids of *Pyrus malus* (the apple) and *Pyrus baccata* (the Siberian crab-apple), and it is probably from this hardy species, spread over Asia and northern China, that all crab-apples first derive. The wild crab-apple *(Malus pumila)* is especially significant as being the parent species of all domestic apples, the varieties of which must certainly be more than the 3,000 which have been named. But the fruit of the crab, whether the bright cherry-red Siberian crab, the conical gold John Downie variety, the yellow Golden Hornet, or the pink-flushed ochre fruit of Transcendant, not dissimilar to the ordinary crab *(Malus sylvestris)* with its greenish diminutive apples, has only a few distinctive uses in the kitchen. Primarily, there is crab-apple jelly, ranging through shades of pink to yellow; verjuice, for centuries the culinary ancestor of vinegar or lemon juice; pickles, and a famous drink, lamb's wool. Raw crab-apples are inedible, although an occasional sharp fruit whose tree has known some earlier domestic strain can be eaten *à la rigueur*. The best use for all wildings, windfalls and crabs (or scroggs, scrabs, wharres or whatever local name attaches to this familiar fruit) is to be gathered for jelly.

As verjuice has some historical rôle in cookery – and could still have its use – it is worth transcribing an old Sussex recipe: 'Take crabs as soon as the kernels turn black, lay them in a heap to sweat, then pick out the stalks and any rotten that may be in them, force them to a mesh [mash], squeeze them in a hair bag, put the juice into bottles and stop them for use.' Verjuice needs to stand at least a month before use; it is a sort of sharp cider, but not a vinegar, with its own excellence, and it was sometimes distilled for pickling. Its rapid disappearance from country life must have been entirely due to the new accessibility of lemons in the nineteenth century, and their ease of use compared with the heavy seasonal labour of crushing whole troughs of crab-apples with wooden mallets and expressing their juice into hogsheads.

CRAB-APPLE JELLY

Wash 5 lb ($2\frac{1}{2}$ kg) apples, and cut them in half to check the soundness of the core. Do not peel them. Place them in a preserving pan with enough water to cover. Boil and simmer for about 45 minutes, until the fruit is pulpy. Strain overnight through a muslin jelly bag, measure the juice that results, and allow 1 lb (450 g) sugar for each pint (600 ml). Dissolve sugar over a low heat, then boil, stirring well, until jelly will set (see jam section, p. 145). Lastly add the strained juice of 2 lemons.

And here is an Irish country version of the preserve, in which the petals would originally have been those of the wild dog-rose. Petals of older varieties of cultwater rose could be used, but those of modern varieties tend to be bitter.

CRAB-APPLE AND ROSE PETAL JELLY

Cut 3 lb ($1\frac{1}{2}$ kg) crab-apples into coarse pieces, without peeling or coring them, cover with water and boil until reduced to a pulp.

Strain this, but do not squeeze, through a jelly bag to obtain a juice, and to each pint (600 ml) of this allow 1 lb (450 g) sugar. Dissolve sugar over a low heat, then boil the sugar and crab-apple juice with 2 handfuls rose petals until the liquid sets when a little of it is poured on a plate. The boiling may take about 20 minutes, but test frequently for a set. Pour the jelly into jars and cover them before the jelly grows cold.

A traditional country recipe gives this method of making pickle:

'To every gallon of white vinegar add one ounce (25 g) of common salt, one ounce (25 g) of mustard seed, half an ounce (15 g) each of mace, whole pepper and cloves, two capsicums, two bay leaves and one drachm [i.e. an eighth of an ounce (3 g)] of cayenne. Boil for twenty minutes and when cool pour over the crab-apples, which have previously been washed in salt water, and let dry before putting into jars.'

Finally, the cordial drink, widely known in Elizabethan times, which Pepys drank late into the night, and which was still made in country districts up to the end of the nineteenth century. Wherry, a more potent rustic liquor, was made from the pulp of crab-apples after the juice had been extracted. The apples were roasted in the hot coals of open fires.

SUSSEX LAMB'S WOOL

Take 8 roasted crab-apples, mash them, and add to them 2 pints (1·2 litres) good old ale. Press through a strainer, add $\frac{1}{2}$ teaspoon each of grated ginger and nutmeg, sweeten to taste, make very hot – but do not let it boil – and drink it hot.

NAMES: Fr. *pomme sauvage, pomme d'apis;* Ger. *Holzapfel*; It. *mela selvatica*.

STORAGE: As apples, but preferably crab-apples should be used as soon as available.

Cranberry

THIS shrub, with its round bright crimson berries *(Vaccinium oxycoccus* and in Britain *V. palustris)*, is the source of fruit more popular in America than Britain, where it grows wild in northern and damp locations. The American cranberry, wild up to the arctic regions, is also cultivated to meet the demand for the traditional sauce that is served with roast turkey and with game. The fruit is too acid to eat raw, and even cooked with sugar presents too sharp a taste for some palates. Nevertheless, the mottled berries have featured in Irish cookery (they are known as bogberries in Ireland), in Welsh tarts and pies, and in numerous jams, jellies and pies in England and America. Their marshy habitat may have accounted for the earlier name of 'craneberry' and 'fenberry'; the mountain cranberry *(V. vitis-idaea)* also appears under the name of cowberry or foxberry. Frozen cranberries are available, and adequate for all cranberry recipes.

CRANBERRY SAUCE

Take 1 lb (450 g) washed cranberries and place them in a deep saucepan with 12 oz (675 g) sugar. Cover with 1 pint (600 ml) water and bring to the boil. Cover the saucepan, reduce the heat and cook slowly for about 10 minutes, or until the skins of the berries burst. Skim the sauce, and leave it to cool. Additions during the cooking of the sauce – the grated rind of a lemon or an orange, or the juice of either, or some spice and a couple of tablespoons of white wine – vary the piquancy.

CRANBERRY ORANGE MOULD

Combine 1¼ lb (550 g) stemmed and carefully selected cranberries, 1 lb (450 g) sugar and ½ pint (300 ml) juice from fresh oranges in a large pan. Sprinkle in the grated rind of half an orange, and gradually bring the mixture to the boil. Let it simmer for 10 minutes, by which time the mixture will have thickened and the berries will have burst. Pour into a 2-pint (1·2-litre) mould and leave to cool, covering the mould after a while and chilling it for several hours. Turn out by inverting on to a cold dish.

RADNOR CRANBERRY TART (TEISEN LYS)

Select 1 lb (450 g) cranberries, free of leaves and bits of stalk, and rinse in cold water. Put the fruit into a basin with ½ lb (225 g) raisins that have been stoned and halved, and sprinkle 1 oz (25 g) sugar all over the fruit. Leave overnight, and the next day turn out the mixture into a small saucepan and bring just to the boil. Simmer for 5 minutes, add ½ teaspoon vanilla essence, and leave on one side to get cold. Next line a plate with rich shortcrust pastry (see p. 11) and spread the fruit on thickly. Cover with pastry and bake for 20–25 minutes in a brisk oven (425°F/220°C/mark 7). Serve the tart either hot or cold, but always accompanied by cheese.

This German cranberry dish can be served with meat, or with dumplings or pancakes:

PREISELBEERSCHÜSSEL

Peel 1 lb (450 g) pears, and cut them up into small pieces. Cook until almost soft in 1½ pints (900 ml) water to which ½ lb (225 g) sugar has been added. Now wash up to 1 lb (450 g) cranberries and, adding a stick of cinnamon, put them with the pears and cook until the berries are soft. Take out the cinnamon, and add 1½ tablespoons vanilla sugar. Eat hot or cold.

SWEDISH CRANBERRY PEARS

Rinse a good 2 lb (1 kg) selected cranberries, and simmer in a saucepan in 2½ pints (1½ litres) cold water until nearly reduced to a pulp. Strain through a cloth or jelly bag, add 1½ lb (675 g) sugar, dissolve sugar, then boil until clear. This syrup should be well skimmed. Then peel about 5 lb (2½ kg) moderate-sized cooking pears, divide lengthways, core, and place in the syrup. Simmer gently in an uncovered pan until tender. Remove the pears to a flat dish to cool, then place in jars that have been rinsed round with brandy. Reduce the syrup a little by boiling and leave to get quite cold. Pour over the pears, and cover the jars with paper. Kept cool, these will remain sound for several weeks. Serve cranberry pears with whipped cream.

CRANBERRY AND APPLE PUDDING

Peel, core and slice 1½ lb (675 g) cooking apples and with these mix ½ lb (225 g) Demerara sugar and 4 oz (100 g) washed and dried cranberries. Put this mixture in a greased basin lined with suet crust (see p. 12). Add a tablespoon or two of water. Cover with suet crust, sealing the edges. Cover with buttered greaseproof paper or foil and steam for 2½ hours. Serve hot, with cream.

NAMES: Fr. *airelle*; Ger. *Moos-*, *Moorbeere*, *Preiselbeere*; It. *mortella di padule*; Sp. *arándaro*.

STORAGE: Keep in a refrigerator in a pack; use within ten days.

Crystallized Fruit

THE preservation of fruits by immersion in boiling sugar (as opposed to pickling in brine) is of great antiquity, and references to the use of candied pieces of lemon or orange, or other fruits, in early recipes are frequent. Methods were not necessarily as subtle as those of today; scarlet barberries, a favoured eighteenth-century candied decoration, were simply dipped in bunches in boiled sugar (syrup), left, removed, and replaced in reboiled sugar at 'candy' point. Some 'candying' seems to have been little more than a sugar glaze.

The process is in fact a simple one, but slow and requiring the assurance of practice and experience. For most fruits, ten days is needed, and for large pieces up to two weeks, so that the candied fruit is covered with sugar not once, but a number of times, with an increasing sugar content in the syrup as the fruit's water is replaced by the sugar. The advantage of patiently candying fruit is that it preserves it for a considerable time, and at little cost – commercially prepared crystallized fruits, those of France and Italy especially, where they are prepared to perfection, are extremely expensive, although prepared mixed candied peel is perhaps an economic purchase for the kitchen.

The first step is to choose only fruits that are firm, e.g. apricots, peaches, plums, pears, stoned cherries; soft fruits will not candy successfully. Peel and quarter the larger fruit, weigh it, stew it, and place in a bowl large enough for it to be covered with the candying syrup. For each 1 lb (450 g) fruit, take $\frac{1}{2}$ pint (300 ml) of the cooking liquid, making up the quantity with water if necessary. To each $\frac{1}{2}$ pint (300 ml) liquid, add 6 oz (175 g) sugar. Dissolve the sugar in the liquid over a low heat, then bring to the boil, and cover the fruit with the hot syrup. Leave for 24 hours. Drain off the syrup, and add 2 oz (50 g) sugar to it. Boil again, never allowing mixture to boil until sugar has dissolved, and pour over the fruit. Repeat for 2 more days, each day adding 2 oz (50 g) sugar. On the next (the fifth) day add 3 oz (75 g) sugar to the drained off syrup. Add the fruit to the syrup, boil for 3 minutes, return the fruit and the syrup to the bowl and leave for 48 hours. Then repeat this process, leaving the fruit this time for 96 hours (4 days). The syrup should now have reached a honey-like texture. Lift out the fruit, spread it on a rack to dry in a warm place (e.g. an oven with residual heat, or an airing cupboard), turning it now and then to even the drying process. Dry, the fruit should be wrapped in waxed or greaseproof paper, packed in wooden boxes, and kept in a cool place.

Crystallized fruit involves a further finishing process, which gives the crystalline sugared exterior: the candied fruit is dipped in boiling water and rolled in granulated sugar. For glacé fruit, with its glazed shiny finish, dip the fruit, already candied, in boiling water for a quarter of a minute and then briefly in boiling syrup; if it clouds, repeat the process until the right translucency is obtained.

Candied fruits, pleasant but filling when eaten alone, also form part of not a few dessert recipes.

TUTTI FRUTTI

Flavour ½ pint (300 ml) double cream with 2 teaspoons each of maraschino, rum and curaçao. Whip it well, and set the cream in little dishes; freeze it, on ice if possible, for at least an hour. Take various crystallized fruits that have been soaked in the liqueurs and cut them into small shreds. Decorate each dish with these to make a wreath round the top before serving.

OMELETTE DE FRUITS GLACÉS

Take the yolks of 6 eggs and beat with a little sugar. Add a pounded macaroon, some crystallized fruits, chopped finely, and 1 tablespoon maraschino. Then whisk 6 egg whites to a stiff froth and fold into the other ingredients. Shape the mixture on a shallow dish, sprinkle with castor sugar and cook for 15 minutes in a moderate oven (350°F / 180°C / mark 4). Serve the omelette as soon as done.

BEIGNETS DE FRUITS GLACÉS

Prepare a batter from ½ lb (225 g) flour, 2 tablespoons melted butter, a pinch each of sugar and salt, 1 whole egg, ½ pint (300 ml) beer, $\frac{1}{2}$ pint (300 ml) water and 1 tablespoon brandy. Put the flour, salt, sugar, butter and egg in a large bowl, mix together, and dilute with the beer and the water, which should be warm. Finally, add the brandy. Let the batter stand in a warm place, and whisk before using. Meanwhile, cut the crystallized fruit – apricots, peaches, pineapple and pears – into halves or quarters, sprinkle with sugar and then with kirsch or brandy, and leave to steep for at least half an hour. Drain the fruit when ready to proceed, and dry it. Dip each segment into the batter and deep-fry in fat which should be smoking hot. Drain again, dry in a cloth and place on a shallow baking dish. Sprinkle with castor sugar and glaze under as hot a grill as possible.

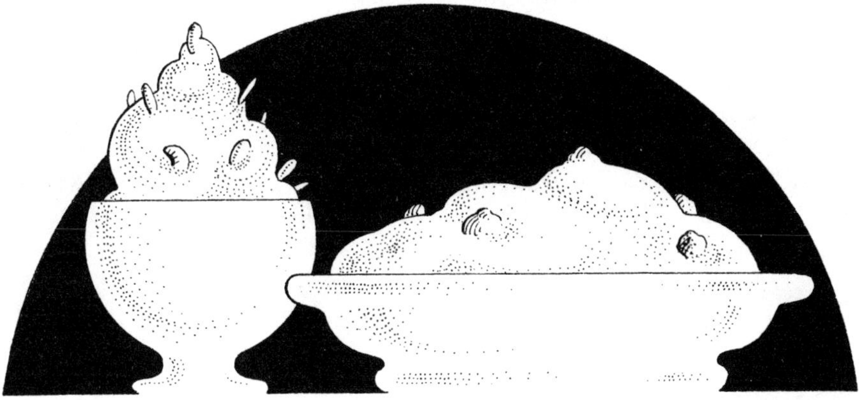

DIPLOMATE À LA CRÈME

Take a number of sponge fingers and spread them with apricot jam. Set them out in layers in a buttered mould. Put in between each layer a few currants, some pieces of angelica, a little candied peel, chopped fairly small, and some glacé cherries. Arrange all this without pressure, as the fingers will enlarge on cooking, and do not fill the mould completely. Pour over it all a *crème* that will fill all the empty spaces: the *crème* is prepared from 1 pint (600 ml) milk, heated almost to boiling point, sweetened and left to cool, and the beaten yolks of 4 eggs, well mixed into the milk; strain before filling the mould. Stand the dish in boiling water and cook for about 45 minutes in a moderate oven (325°F/165°C/mark 3). Serve quite cold, turned out with a *crème* all round it, this being cooked separately. Make this from 1 pint (600 ml) milk, warmed, a little sugar and 3 eggs; pour the milk on to the eggs, strain into the top of a double boiler and stir over a gentle heat until the sauce thickens.

This very agreeable sweet has an uncooked version, in which the fingers are soaked in equal parts of water, rum and kirsch, and layered in the mould with mashed strawberries and raspberries between the layers. The mould is weighted and left in a refrigerator for 24 hours. The pudding is then turned out and served with a *crème* as for the cooked *diplomate*.

This Cambridge recipe dates from the eighteenth century:

THE DEAN'S CREAM

Spread a number of sponge cakes with raspberry jam, and a similar number with apricot jam, and arrange them alternately in a dish. Add some ratafias (small macaroons), and soak the cake with sherry. Sweeten $\frac{3}{4}$ pint (450 ml) double cream with castor sugar, add 4 tablespoons brandy and whip until very thick. Pour this over the soaked cake. Finally, decorate the pudding with glacé cherries, pieces of crystallized pineapple and candied angelica stems.

This Russian dessert cake was traditionally eaten, as the name implies, at Easter, at the end of the Lenten fast:

PASHKA

Put 6 oz (175 g) seedless raisins in hot water to soak, with 1 tablespoon sugar. Next, chop into tiny slivers 5 tablespoons various candied fruits: angelica, candied peel, crystallized ginger, cherries and orange, etc. The basis of this dish is 1 lb (450 g) fresh, unsalted cream cheese into which you beat $\frac{1}{2}$ pint (300 ml) sour cream. While beating this work in 5 oz (150 g) icing sugar, $\frac{1}{2}$ teaspoon vanilla essence and $\frac{1}{2}$ teaspoon almond essence. After 30 minutes the raisins will have swollen; drain and stir into the cheese mixture, then add the chopped crystallized fruit. Mix again, and form (in a mould if available) a stiff pyramid shape, strengthening the *pashka* with more sugar or cheese if necessary. Cool overnight, and before serving decorate with walnut halves, blanched almonds and crystallized cherries. (The Russian decoration includes the letters XB – Christ is risen – in various candied pieces.)

Date

THE date, fruit of a feather-leaved palm *(Phoenix dactylifera)* of the arid regions of south-western Asia and northern Africa, has hundreds of named varieties, some merely local. It has also been introduced to the New World, and has been cultivated in the United States since the latter part of the nineteenth century. The ordinary distinction is simply between soft, sweet dates of the kind exported and the hard, dry, long-keeping dates gleaned by the Arabs; a third sort of fibrous date has no keeping qualities and little sugar. Recorded as being in cultivation as early as 3,000 BC, the date is a valuable and staple food in North African countries. A reliable and respected tree, the palm bears fruit when four years old and continues to do so for about seventy years; it is also a source of useful fibres, wood, and the sap on which palm wine is based. The ovoid yellowish fruit, with a grooved nut, hangs in bunches of some forty strands, with about thirty dates on each strand; four or five bunches ripen at a time, and the tree produces about 100 lb (50 kg) fruit a year.

The golden-brown sweet dates and the dry (Deglet Nour are the most generally used, and Tafilats on the stem used to be the preferred dessert date) are both highly nutritious, with a large proportion of sugar, and valuable vitamins and minerals. Some loss takes place in the process of preservation, but the date remains a rich dietetic food, fully appreciated in the Arab world. Yet the date has never entrenched itself in Western eating habits. It occurs in odd recipes from Roman and mediaeval times, as in an Elizabethan recipe for pike boiled in wine with dates, oranges and spices, but has little presence until Mrs Beeton's day, appearing then in puddings, 'shapes', suet rolls and nursery foods.

Cakes and breads including dates are pleasant and substantial, and can be varied by the inclusion of walnuts, a nut that goes well with the date. But a certain stodginess is inevitable where dates are used (the Tunisian *makroudh* – date squares – is a particularly dense sweetmeat), and perhaps for this reason the French rarely make use of them except in *dattes fourrées*, where the stuffing is almond paste flavoured with kirsch and the date is rolled in icing sugar. As a single date has the calorific value of a small steak, one may be enough, and a dish of filled dates – with chives and cream cheese, broken walnuts, Brazil nuts, crystallized ginger or pineapple pieces – makes a fitting dessert.

DATE AND LENTIL PILAF

Chop 1 small onion finely, and add to it 4 oz (100 g) chopped dried apricots; mix in 2 oz (50 g) chopped blanched almonds, 2 oz (50 g) sultanas and 6 oz (175 g) chopped stoned dates. Fry this mixture lightly in butter, and set on one side. Cook 1 lb (450 g) lentils until tender, and, separately, 2 lb (1 kg) rice until almost ready. Prepare butter-water by heating 4 oz (100 g) butter with ¾ pint (450 ml) water and bringing to the boil. Put a layer of rice in the pilaf pan, then the lentils, another layer of rice, and the fruit and onion mixture. Cover with a final layer of rice, then add the butter-water. Cover the pan and steam for half an hour over a very low heat; the pan must be tightly closed. Serve on a large serving dish.

DATE WHIP

Stone 14 oz (400 g) dates, and chop finely. Chop 4 oz (100 g) shelled walnuts. Whip 1 pint (600 ml) double cream until stiff, and fold in 4 oz (100 g) castor sugar and a little vanilla essence. Add the nuts, then the dates. Chill thoroughly before serving.

MARZIPAN-STUFFED DATES

Take a number of firm, ripe dates – preferably fresh – and slit them lengthwise, making a clean cut not more than halfway round. Remove the stones. Break 4 eggs into a saucepan and add 5 oz (150 g) castor sugar and 5 oz (150 g) flour. Mix this thoroughly, then stir in 1 pint (600 ml) boiling milk that has been flavoured with a vanilla pod. Continue stirring over a gentle heat until the mixture has the texture of cream, then let it cool. Now add 4 oz (100 g) melted butter, 4 oz (100 g) ground almonds and 12 oz (350 g) castor sugar. Stir again, very thoroughly, and let the mixture get quite cold. The dates can then be stuffed with marzipan; they should be eaten within a day or two.

DATE PUDDING

Place in a mixing bowl 4 oz (100 g) stale white breadcrumbs, 6 oz (175 g) stoned dates, chopped up finely, and 4 oz (100 g) suet. Mix them all together. Add 2 tablespoons treacle. Mix again. Add the beaten yolks of 2 eggs, and then the 2 egg whites, whisked until stiff. Put the whole mixture into a buttered mould, cover with buttered greaseproof paper or foil, and steam the pudding for 2 hours. Turn out, and serve quite hot with rum-flavoured custard.

SUDANESE DATE PUDDING

Take 10 oz (275 g) dates, stone them, and cut them up into quarters. Boil in $1\frac{1}{2}$ pints (900 ml) water, adding 1 teaspoon baking powder, a pinch of salt and 3 oz (75 g) sugar. Continue boiling for 20 minutes. Mix $1\frac{1}{2}$ oz (40 g) flour with a little water, then gradually add more water to make the quantity up to $\frac{1}{2}$ pint (300 ml); stir this gradually into the date mixture. Continue cooking until thick, then use an electric blender to ensure a smooth and thorough mix. Serve hot in small glass dishes, with a pat of butter melting on top of each.

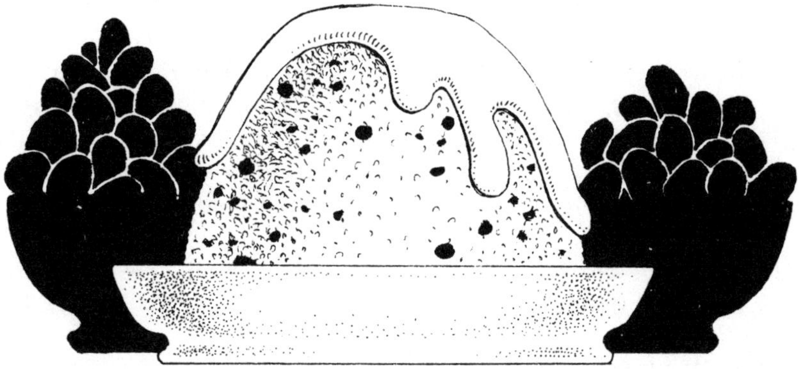

NAMES: Fr. *datte*; Ger. *Dattel*; Sp. *datil*; Port. *tamara*.
STORAGE: If well wrapped, dates can be refrigerated, but loss of moisture takes place in time.

Dried Fruit

THE drying of fruits – and there are not many for which this is an advantageous method of preservation – cannot be carried out in climates less warm than Mediterranean or sub-tropical. Sun-drying is, naturally, the best way, and the fruits that are most successfully so modified are grapes (into raisins, sultanas and currants), apples, apricots, dates, figs, pears, peaches and plums (into prunes). In England, apples were once commonly dried in ovens, the hard, firm-skinned Norfolk biffins being particularly suitable for this, or they were peeled and cored and strung up whole in lines to dry hanging from the ceiling of a warm room. Once dried, fruit has a new character, which it retains after soaking or stewing. Naturally, dried fruits find their way into a number of recipes, their use in cakes and puddings being valuable where it is necessary to keep the water content to the minimum. They may also be eaten on their own, dry, but a minimal addition of some liquid is a great improvement: dried apricots, for instance, provide an attractive dessert with no more preparation than covering them with brown sugar and baking slowly in the oven.

This Silesian recipe, 'The Kingdom of Heaven', relies substantially on dried fruit:

SCHLESISCHES HIMMELREICH

Soak ½ lb (225 g) mixed dried fruit overnight, using apricots, pears, prunes and apples. Partly boil the fruit and strain. Then take

4 pork chops and simmer gently in the fruit juice until tender. Add the fruit to the chops and cook until the fruit softens. Using another saucepan, heat 1 oz (25 g) butter, add 1 oz (25 g) flour when it melts, and stir for 2 minutes while it fries. Add the liquid in which the chops and fruit have been cooking, stir to make a smooth sauce, then add the chops and the fruit. Allow a few more minutes' cooking. The dish is usually served with potato dumplings.

QUATRE MENDIANTS

A dessert of long standing in France, this dish simply consists of a plate of dried figs, raisins, filberts (*avelines*, hazelnuts) and blanched almonds. It is a reminder that fresh fruit was not always available all the year round. The mendicant orders represented are the black friars (Dominicans, raisins), the white friars (Carmelites, almonds), the grey friars (Franciscans, figs), and the Austin friars (Augustines, filberts).

STUFFED DRIED FRUIT

Wash ½ lb (225 g) prunes, a number of dates and ½ lb (225 g) dried whole apricots and steam until they are soft. Let them cool. Cut a slit in each piece of fruit, remove the stone, and fill with unsalted ricotta cheese, garnishing each with half a walnut.

BAKED STUFFED APPLE

Take 1 very large cooking apple per person and slice off the top horizontally, keeping the stalk. Set this lid aside, and core the apple, being careful not to pierce the base. After removing the core, hollow out the centre a little more to take the stuffing. Now take raisins, chopped dates, chopped dried figs and sultanas, all of which have soaked for about half an hour in a mixture of water and brandy, and fill the hollow centre. Pack in, not too tightly, and pour 1 teaspoon honey down the centre. Place a knob of butter on top and replace the lid. Stand in a baking pan in a little water. Slit the skin of the apple very slightly in four places, and butter lightly before baking in a moderate oven (350°F/180°C/mark 4) for about an hour. When the apple foams through the slit skin, it is ready to eat. Serve with cream and a sprinkling of kirsch.

SIR ROBERT WALPOLE'S DUMPLINGS

Shred 1 lb (450 g) suet into small pieces, and grate 12 oz (350 g) stale bread. Clean 4 oz (100 g) each of currants, sultanas and raisins, and cut up into very small pieces 2 oz (50 g) each of orange peel and citron peel. Mix all these together, then add a pinch of cinnamon and 4 oz (100 g) castor sugar. Work in the yolks of 6 eggs and the whites of 3 of them, and beat together until a very stiff paste is achieved. Make dumplings the size of apples from the mixture and tie them up in pieces of clean cloth. Put them in a pan of boiling water, and cook for approximately 1 hour. Serve hot, pouring over them a sauce of melted butter, wine and sugar.

Drinks

FRUIT-BASED drinks, whether alcoholic or soft, need no apology. Between the venerable and arcane processes that transform the grape into wine, and the simple extraction of juice from fresh fruit for drinking without delay, lies a whole area of agreeable concoctions. It is true that the housewife's stillroom, with brews based on the hedgerow's yield, is a thing of the past, but many good cups and cordials, and homemade wines, will repay the effort involved. As for soft drinks, they demand to be consumed soon after making, as in any fruit drink fermentation may occur if it is neglected, but summer drinks of fruit juice, garnished with slices of orange and lemon, and generously heaped with crushed ice, are not likely to last long.

APPLE DRINK

Wash 2 lb (1 kg) apples and, without peeling them, cut them up and core them. Place them in a pan, and add 1 oz (25 g) cinnamon stick and the grated rind of a lemon. Add the juice of 2 lemons, 3 pints ($1\frac{3}{4}$ litres) water, and boil until soft. Mash and strain the liquid through a sieve, add 4 oz (100 g) sugar to it and bring to the boil again for a few minutes. Leave to cool, and serve iced.

BLACKCURRANT CORDIAL

Put blackcurrants and Demerara sugar in alternate layers in a large earthenware jar, using an equal weight of fruit and sugar. Crush 12 cloves, place them in a muslin bag, put them in the jar, and seal it so that it is completely airtight. Leave for 6 weeks, then strain off the syrup. For each 6 lb (3 kg) weight of berries and sugar originally used, add 5 tablespoons brandy to the syrup. Stir well. Bottle the cordial, and seal with tight corks.

CHERRY PUNCH

Stone 1 lb (450 g) cherries and mash them up. Put them in a bowl with the juice of 3 lemons and 2 oranges, and 2 slices of pineapple.

Cover with sugar and leave for an hour, then press and strain. Heat up the liquid, then let it cool. Finally, add 2 pints (1.2 litres) each of soda water and claret, and 4 oz (100 g) ripe, stoned cherries.

EAU DE FRAISES

Take 1 lb (450 g) ripe, perfect strawberries. In an enamelled saucepan put 3½ pints (2.1 litres) water and 12 oz (350 g) lump sugar, and let the sugar dissolve over a very low flame. Then bring the liquid quickly to the boil, remove from the flame and put in the strawberries. Cover, and let the strawberries infuse for an hour. Then strain the liquid through a fine cloth, and keep it on ice. Before pouring out to drink, moisten the glass with a few drops of kirsch.

ELDERBERRY WINE

The berries of the elder tree are best gathered on a dry day; clean them from their stalks and place in an earthenware pan. For every 15 lb (7½ kg) berries, pour over 2 gallons (9½ litres) boiling water, pressing the berries into the water. Cover closely and leave for 24 hours. Then strain the juice from the fruit through a sieve, squeezing the berries for additional juice. Measure the juice: to every gallon (4¾ litres) add 3 lb (1½ kg) sugar, ¾ oz (20 g) cloves, and 1 oz (25 g) root ginger. Boil for 20 minutes. Remove the scum as it forms. Allow to cool, and strain it into a clean, dry cask, allowing room for the yeast to work. Pour into the bung-hole with care 1 tablespoon fresh yeast mixed with a very small amount of the liquid. The wine should be ready in about 1 month, but should be kept for about 3 months before drinking.

FRUIT CUP

First boil ½ pint (300 ml) pineapple juice with 4 oz (100 g) sugar for 10 minutes. Prepare the fruit by cutting it up into small pieces, disposing of skin, pith and seeds. Use 1 grapefruit, 1 apple, 1 orange, 1 banana, 4 oz (100 g) pineapple and 4 oz (100 g) grapes. Pour the heated juice over this fruit, add the juice squeezed from 2 lemons, and leave for 12 hours. Add 3 parts iced water to one of fruit juice, retaining the fruit pieces. Serve very cold.

The prolific dark-red berries of the hawthorn provide this Yorkshire recipe:

HAWBERRY WINE

Put 5 lb ($2\frac{1}{2}$ kg) ripe hips and haws in a pan with 1 gallon ($4\frac{3}{4}$ litres) boiling water. Stir every day for a fortnight to ferment. Then drain off the liquid, pressing the berries to obtain more, and boil it for 15 minutes. Then strain and measure it, adding, to each gallon ($4\frac{3}{4}$ litres) liquid, 3 lb ($1\frac{1}{2}$ kg) sugar, 1 oz (25 g) whole root ginger, bruised, and 1 tablespoon cloves. Boil all this yet again for 15 minutes, then strain and bottle when it has grown cold.

KIRSCH PUNCH

Mix together $\frac{1}{4}$ pint (150 ml) kirsch and 1 pint (600 ml) fresh orange juice. Add $\frac{1}{2}$ pint (300 ml) lemon juice. Stir in 1 tablespoon pineapple syrup, and add castor sugar to taste. Chill the punch thoroughly, and add 12 whole strawberries, ripe, of the same size and hulled.

This restorative makes full use of the astringent qualities of the lemon:

LEMON SCORCHER

Squeeze the juice of half a lemon into a wineglass, and add 1 tablespoon old brandy, or Hollands gin, and a dusting of cayenne pepper. Mix well. Add further lemon juice if needed. Strain out any lemon pips.

LEMONADE

Pare 2 lemons as thinly as possible, and put the rind in a jug. Add 1 oz (25 g) castor sugar, or more to taste. Pour over the rind 1 pint (600 ml) boiling water and cover the jug, leaving it until it is cold. Now add the juice of the 2 lemons. Strain the lemonade before serving. Double the quantities if required, but add the juice of a fifth lemon.

LOGANBERRY PUNCH

In a pan put ½ lb (225 g) loganberries and 4 oz (100 g) sugar. Place over a low flame, and when the juice begins to run crush the loganberries with a wooden spoon against the sides of the pan. Remove from the heat when the juice has been extracted, strain it into a jug and set aside to cool. Next have ready the juice of 2 lemons and 3 oranges, both strained, and 2 pints (1·2 litres) cold water. Pour all the fruit juices and the water into a punch bowl and stir, adding lumps of ice. Before serving, decorate with thin slices of banana, a few whole strawberries and some whole loganberries.

ORANGE PUNCH

Prepare 3 pints (1¾ litres) fresh orange juice, strained. Heat gently with 6 oz (175 g) sugar. Brew enough Indian tea for 6 cups (about 1½ pints (900 ml)), strain out any leaves, and add the juice of 2 large lemons. Peel enough oranges to provide at least 12 orange segments, and insert a clove in each. Combine the tea and orange juice and place the orange segments in this. Stir in ½ pint (300 ml) peach brandy. Chill thoroughly before serving.

PEACH AND GINGER DRINK

Take 4 medium sized peaches, fully ripe, and peel and sieve them. Mix in 4 oz (100 g) castor sugar, ½ teaspoon ground ginger, a pinch of nutmeg, and the juice of 1½ lemons. Chill the mixture, then add

2 pints (1.2 litres) ginger beer. Peel and slice an orange into rounds and garnish the drink with these. Serve very cold.

RAISIN WINE

To 1 gallon (4¾ litres) water allow 8 lb (4 kg) raisins. First boil the water and let it get cold, then fill a cask with the water and the raisins, pressing the raisins down well. Partly close the bung, and stir every day, keeping the cask full by adding cold boiled water. Fermentation will be over in 6–7 weeks; now press in the bung and leave for 12 months. Draw off into a clean cask, and fine the wine with isinglass tied in a muslin bag and suspended in it. This wine will keep for years.

STRAWBERRY SHERBET

Mash up 1 lb (450 g) wild strawberries – or, failing these, 1 lb (450 g) small cultivated strawberries – add several slices of fresh lemon, from which the pips have been removed, 1 teaspoon orange-flower water, and ½ pint (300 ml) water. Cover. Leave for 3 hours, then strain through a cloth, forcing the liquid through. Stir in ½ lb (225 g) sugar until it is dissolved, and strain again. Keep the drink refrigerated.

Fig

FICUS CARICA, the fig tree, is a member of the mulberry family, with distinctive lobed palmate leaves and pear-shaped fruit – green, brown or purple – and a thin to moderately tough skin covering a fleshy receptacle in which, uniquely, the flowers of the fig come to perfection and seed. Unless dried or preserved, the fig is not the best of travellers; it would be ideal to eat figs where they grow, in Greece, Asia Minor, North Africa, Provence or Italy. Wall-trained fig trees will, of course, produce fruit in cooler climates than these, as excellent crops in the south of England confirm, but in general the kitchen must rely on well-packed figs – not those kneaded into a different shape to facilitate packing – imported from Malta or Smyrna.

Probably originating in Asia Minor, the fig has been respected from the earliest times: its importance in Hellenic culture and economic life is second only to that of the grape and the olive. Attic figs (could the succulence of these really have persuaded Xerxes to invade Greece?) were introduced into Roman Italy well before the second century BC, for Cato mentions six varieties; Pliny, about AD 77, records twenty-nine sorts. In the first half of the sixteenth century, the fig was brought to England by Cardinal Pole, a few years before Cortez introduced the tree to Mexico. Fig trees reached North America about 1790.

Apart from puddings and a few other cooked uses, the fig provides useful types of syrup, as listed in the *Pharmacopoeia*, makes an excellent jam (for which green figs are more appropriate), has been roasted as a substitute for coffee, and makes a pleasing addition to many salads. Its simplest and most effective appearance is perhaps in an austere *hors d'oeuvre* such as *prosciutto coi fichi*: a perfectly ripe fig from the tree served with thin slices of Parma ham. Parkinson, in 1629, was of like opinion over the recent fig trees 'nursed up in our country', that 'Figges are served to the table with Raysins of the Sunne, and blanched Almonds, for a Lenten dish'.

This is an Egyptian casserole of fresh figs and *bamiya* (okra or Ladies' Fingers), seed pods that have been a favoured vegetable along the Nile for centuries:

FIGS AND BAMIYA

Put 1 lb (450 g) fresh okra in lightly salted boiling water and simmer for half an hour until soft. Take them out, drain, and arrange with 8 figs in a casserole. Add 2 oz (50 g) butter, and sprinkle with black pepper and a little salt. Cook in a moderate oven (350°F/180°C/mark 4) for about 15 minutes. Then sprinkle with brown sugar and cook under the grill until the sugar caramelizes.

Serve with melted butter, as a vegetable dish on its own, with poached eggs, or to accompany a meat dish.

SALADE DE FIGUES AUX LIQUEURS

Take 6 fresh green figs, peel them, and cut into quarters. Arrange in a bowl, cover with castor sugar, and sprinkle on a generous flavouring of brandy mixed with equal amount of orange curaçao. Stir well. Let the fruit cool in a refrigerator. Serve with whipped cream, sweetened, and flavoured with a few drops of curaçao.

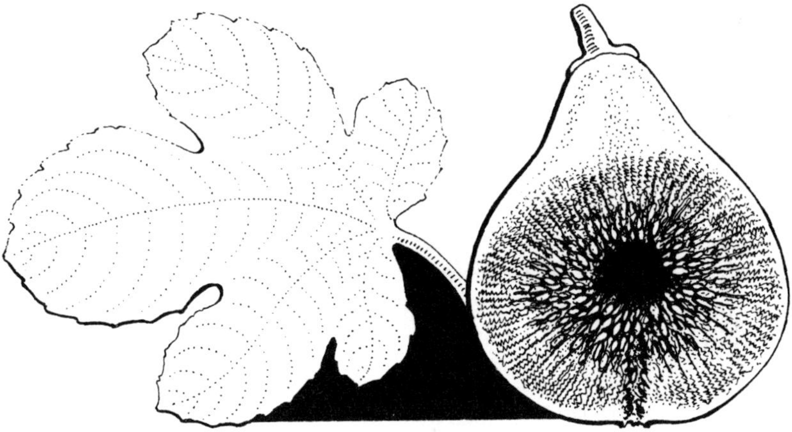

BAKED FIGS

Wash, but do not peel, 6 ripe figs, and stand them upright in an oven casserole. Add a little water and dust the figs with brown sugar. Bake in a very slow oven (275°F/135°C/mark 1) basting often, and pricking each fruit with a fork to release steam. When cooked the figs will be very tender; leave them in the dish, and dust with a mixture of ground cinnamon and ground cloves. Add 5 tablespoons kirsch or Grand Marnier. Chill nearly to freezing point before serving, then drain off the liquid and use it as a sauce. Serve with this sauce and with whipped cream flavoured with the liqueur already used.

STUFFED FIGS

Any number of dried figs can be used: wash, and cut them in half. Fill the centres with a mixture of cream cheese and finely shredded almonds, flavoured with a sprinkling of vanilla sugar. Impale the figs with sticks to hold them together. Dust with castor sugar, or a mixture of cinnamon and castor sugar.

FIGUES FLAMBÉES SABAYON

Using either a bottle of figs in syrup, drained well, or fresh figs, place figs in a soufflé dish to about three-quarters full. Warm slowly in the oven, and just before they are ready to serve pour warmed brandy over the figs and light it. The sauce that goes with the figs is made of 3 egg yolks, 2 tablespoons Marsala, 2 tablespoons white wine, 2 oz (50 g) castor sugar and the rind of quarter of a lemon, finely grated. Place all these in a basin, and place this over the heat in a bain-marie containing hot water. Beat gently until the sauce rises to three or four times its original quantity, and continue beating until it thickens. Take care not to allow it to boil. Serve immediately.

According to Eliza Acton, this fig pudding occurs in the Greek historian's works, with honey taking the place of sugar, wine or sherry, and a boiling time of 14 hours. This recipe is for half the quantity, but uses as many figs as the larger pudding, and adds an egg and a sauce:

HERODOTUS' PUDDING

Mix together ½ lb (225 g) stoned raisins, ½ lb (225 g) beef suet, 4 oz (100 g) breadcrumbs, 3 teaspoons allspice, 1 egg, 1 tablespoon moist brown sugar, the grated rind of half a lemon, 4 dried figs, chopped very finely, and 4 tablespoons sherry. Beat together until smooth, turn into a buttered mould, cover with buttered greaseproof paper or foil, and boil for up to 7 hours. Meanwhile make a sauce of melted butter, to which add brown sugar, grated lemon peel and 8 tablespoons Madeira, and serve it as hot as possible with the fig pudding.

MRS HART'S FIG PUDDING

Chop ½ lb (225 g) dried figs into very fine pieces, and chop 6 oz (175 g) suet into equally small pieces. Mix both these with ½ lb (225 g) breadcrumbs, ½ lb (225 g) moist brown sugar, and half a grated nutmeg. Have 2 well-beaten eggs ready, and fold these in. When the mixture has been well combined, put into a buttered mould and cover with greased paper or foil. Steam for 4½ hours before turning out. Serve hot, with an apricot sauce (see p. 00).

NAMES: Fr. *figue*; Ger. *Feige*; It. *fico*; Lat. *ficus*; Gk. *sykon*.
STORAGE: In a cool place, but immediate use is advised, as the fig deteriorates very quickly.

Fools

FRUIT fools belong to the grand Georgian age of milk and cream in English cookery – there is scarcely any Continental equivalent – when dairy products dominated whole aspects of eating. Their use in desserts can be traced to mediaeval junket (rennet-curdled cream with rose water and sugar), a Norman French innovation; flummeries, trifles, fools, syllabubs and cheese cakes followed in the seventeenth century, and ice cream in the eighteenth. Essentially, a fool is one soft fruit pulped or sieved and whipped into thick cream. Some local variations were known. For instance, Norfolk fool was cream seasoned with eggs and boiled (with no fruit), while Devonshire 'whitepot' was cream, eggs and currants, baked. As might be expected, Victorian cookery elaborated on the simple formula, adding sherry or rum, and serving fools with such biscuits as *langues de chat* or *bâtons à la canelle.* Originally not so much a method of using fruit as an outlet for an abundance of cream, the cooling and delicious fool keeps its place among English desserts.

APPLE FOOL

Cut up and core 1 lb (450 g) apples, leaving the pieces in their skins. Stew, with sugar, until soft. Pass through a sieve and whip in $\frac{1}{2}$ pint (300 ml) double cream. Stir in 1 teaspoon rum; chop and pound a small quantity of candied peel and mix it in. Place in a glass bowl and chill. Serve with fingers of sponge cake, dried and then browned under a grill.

BANANA AND RHUBARB FOOL

Cook some sticks of rhubarb. Blend or sieve to make a purée and sweeten to taste. Peel 6 small ripe bananas and rub through a sieve into $\frac{1}{2}$ pint (300 ml) of the purée. Add $\frac{1}{4}$ pint (150 ml) double cream, stir, and colour delicately with a few drops of cochineal. Serve in glasses. Embellish the top with stiff whipped cream and a few very thin slices of banana.

BLACKBERRY FOOL

Wash some blackberries. Cook them slowly in a pan, with a very little water, for a few minutes. Sprinkle them with sugar and cook again for a few minutes. Pass them through a fine sieve. Cook some apples and pass them through a sieve. Whip cream and gently mix this with the blackberries and the apples together in the proportions of $\frac{1}{2}$ pint (300 ml) cream to 1–1$\frac{1}{2}$ pints (600–900 ml) of fruit purée. Serve from a glass bowl, with sponge fingers.

GOOSEBERRY FOOL

Top and tail 1$\frac{1}{2}$ lb (675 g) gooseberries that are tart rather than sweet, place in cold water in a saucepan with 4 oz (100 g) sugar and cook until they are soft. Press them through a sieve, leave until they are cold, then chill. Whip into the gooseberry pulp 1 pint (600 ml) thick cream, which has been whipped until stiff and itself chilled. Serve in custard glasses.

NORTHAMPTONSHIRE GOOSEBERRY FOOL

Trim the two ends of the gooseberries cleanly, and drop them into a little boiling water until the fruit breaks. Drain off the water, and when the gooseberries are cold again mash them with a fork (do not pass through a sieve), sweeten as necessary with castor sugar, and add an equal quantity of rich cream.

GUAVA FOOL

Drain a 14 oz (396 g) tin of guava pears, and pass the fruit through a sieve to remove the seeds. Stir well, and whip in up to $\frac{1}{2}$ pint (300 ml) double cream, as seems desirable. Serve in glass dishes, chilled, with biscuits.

MEDLAR FOOL

Using fully bletted medlars (see p. 174), cook the fruit gently for up to 15 minutes in a little water to which a drop of rum has been added. Rub through a hair sieve, and mix thoroughly with thick cream, 1 part cream to 2 parts medlar purée. Add brown sugar to sweeten. Serve cold, but not frozen.

STRAWBERRY AND RASPBERRY FOOL

Bruise $\frac{3}{4}$ lb (350 g) strawberries and $\frac{3}{4}$ lb (350 g) raspberries, and pass through a fine sieve. Mix them with $\frac{1}{2}$ lb (225 g) castor sugar and 1 tablespoon orange-flower water. Bring to the boil $1\frac{1}{2}$ pints (900 ml) double cream, stir, and continue stirring until it is cold, then beat in the fruit pulp until well mixed. Put the fool into glasses, and decorate with very small whole strawberries. Chill before serving.

Fruit Cakes

THE great catalogue of British cakes – even greater if mention is also made of the *gâteaux* of France, the *Kuchen* of Germany and the equivalents of other nations – has yet to be written. Omitting 'small' cakes, tea-cakes, buns, scones, fruit breads and biscuits, there remains a huge variety, many of which embody some fruit, even if only a trace of almonds or lemon peel, a few currants or raisins.

The origins of cake are lost in the beginnings of cooking, the mediaeval cake being a form of bread enriched with spices (cinnamon, ginger, anise, coriander and caraway seeds) and later with dried fruit, as in the traditional 'plum' cakes. These were often huge cakes, with many pounds of ingredients, and like bread were baked in the bread oven, on greased papers. The smaller, modern cake may be said to have sprung from the invention of the kitchen range by Thomas Robinson, in 1780, and the adoption of the tin hoop to shape the mix at about the same time. This period also saw the infancy of the wedding cake ('bride cake') with its novel double layer of 'almond icing' (marzipan) and sugar icing.

It is indeed largely with ceremonies and occasions that the national British cake has grown, while a second rank of cakes and buns has persistently been associated with places (Shrewsbury, Banbury, Eccles, Chester, etc.) and traditional festivals (Twelfth Night, Mid-Lent Sunday for Simnel cake, etc.). Of the more important cakes it is said that they should be as 'black as the devil, heavy as sin, and sweet as young love'. To achieve this a light hand and a strong arm are needed, a scrupulous attention to ingredients, and the ability to plunge a knife into the heart of a cake to see whether or not it is, as the Scots say, glary. As for fruits, the most suitable are dried, candied or glacé fruits; nuts either in slivers or ground; soft fruits in purée or conserve form for sandwich layers, and citrus fruits in subtle amounts with gratings of peel. Where spirits are mixed in, brandy, rum and whisky are generally interchangeable, as are sherry and Madeira.

Mandelkoek is one of many Dutch cake recipes:

ALMOND CAKE

Blanch and pound 1 oz (25 g) bitter almonds with 1 lb (450 g) sweet almonds, adding not more than 2 teaspoons rose water as the almonds are being ground. Beat together 12 oz (350 g) castor sugar and the yolks of 10 eggs until frothy. Whisk the egg whites until they are stiff, then add small amounts of egg white alternately with the almonds to the egg yolks and sugar. Stir in 4 heaped tablespoons pounded macaroon biscuits. Butter a large cake tin generously and turn the mixture into it. Bake slowly (300°F/150°C/mark 2) for about an hour.

HAZELNUT CAKE

Beat the yolks of 3 eggs and 3½ oz (90 g) sugar for 20 minutes, or beat until thick with an electric mixer, then add 4 oz (100 g) hazelnuts, finely ground, 2 tablespoons breadcrumbs, thoroughly dried, ¼ teaspoon baking powder, and the whites of the eggs, which have been whipped up to a stiff snow. Butter a cake tin and fill it with the mixture. Bake for at least half an hour in a moderate oven (325°F/160°C/mark 3). When cold, ice with coffee icing.

Ranginak, an Iranian date cake, is a speciality of the town of Shiraz, itself a centre for the cultivation of some of the best Iranian dates.

DATE CAKE

Using a cake tin with a detachable bottom, pack in as tightly as possible, standing them on their ends, 1 lb (450 g) stoned, dried dates. Level them down firmly. Heat 4 oz (100 g) butter, and stir into it, a little at a time, 4 oz (100 g) flour, and let this mixture simmer until it turns a golden brown. Now spoon it over the dates, levelling it out and letting it cool. Take 1 oz (25 g) castor sugar and mix into it 1 tablespoon cinnamon; sprinkle this over the cake before serving and after it has been lifted out of its tin.

'Sly' cakes, sometimes called cheats, are English country cakes that conceal their richness, looking plain on the outside:

FIG SLY CAKES

The filling consists of 6 oz (175 g) dried figs, 3 oz (75 g) chopped walnuts, and 1½ oz (40 g) each of currants and raisins; chop all these together, add 2 oz (50 g) sugar and simmer in enough water to cover the mixture. When the fruit is tender and pulpy, let it cool. Make the pastry with ½ lb (225 g) flour, 2 oz (50 g) castor sugar and a pinch of salt, lightly rubbing in 3 oz (75 g) each of lard and butter. Chill, roll out, and line a rectangular baking tin. Spread in the fruit filling and cover with pastry, sealing the edges and marking it into squares. Brush with egg yolk beaten with milk. Bake in a hot oven (400°F/200°C/mark 7) for 15–20 minutes. Cut the cake up into squares when cool.

LONGFORD CAKES

Mix together 1 tablespoon each of ground almonds, chopped apple and chopped currants and the grated rind of 1 lemon, with sufficient apricot jam to bind them. Put a little of this mixture between two pieces of puff pastry about 2½ inches (6 cm) across (see p. 12), press the edges together firmly, and brush over with water. Finally, place on a baking tray, sprinkle with castor sugar and bake in a quick oven (425°F/220°C/mark 7) for 15–20 minutes.

WALNUT CAKE

Beat 4 oz (100 g) butter thoroughly, and mix in 6 oz (175 g) castor sugar, 3 eggs and 4 oz (100 g) plain flour. Add ½ teaspoon baking powder, and lastly 6 oz (175 g) chopped walnuts. Bake in a large, shallow cake tin, in a moderate oven (350°F/180°C/mark 4), for 20–30 minutes. When cold slice in half and spread a layer of raspberry jam between the halves. Then ice the cake, and decorate with angelica and walnut halves.

GÂTEAU BIGARREAU

Break 3 eggs into a basin and gradually whisk in 4½ oz (115 g) castor sugar; stand the bowl over hot water and continue whisking until the mixture thickens and turns white. Fold in 3 oz (75 g) plain flour. Butter a large, shallow cake tin, dust it with flour, and pour the mixture into it. Bake in a moderate oven (350°F/180°C/mark 4) for 20–30 minutes. Meanwhile, put 3 tablespoons apricot jam and 2 tablespoons redcurrant jelly with not more than ¼ pint (150 ml) water in a pan and simmer. When this attains the consistency of cream, strain and leave to cool. Next, stone 1 lb (450 g) whiteheart cherries and set to one side. Take a ¼ pint (150 ml) cream, whip it, sweeten with castor sugar, and fold in 2 tablespoons *pralin* (made by dissolving 3 oz (75 g) sugar with 3 oz (75 g) unblanched almonds, caramelizing them, and grinding to powder when cold). Split the cake horizontally, and sandwich the cream between the two halves. Then brush the top and sides with the jam and jelly glaze. Arrange the cherries over the top surface. Brush these over again with glaze, more thickly, and brush the sides again, pressing in either browned, broken nuts or more *pralin*.

RICH FRUIT CAKE

Cream ½ lb (225 g) butter, and add to it ½ lb (225 g) castor sugar. Cream it again, and, one at a time, add 4 eggs, thoroughly beating the mixture between each one. Blend in 3 oz (75 g) flour as lightly as possible. Gently mix in 1 lb (450 g) mixed dried fruit – currants, raisins and sultanas – and 6 oz (175 g) candied peel, chopped finely; mix in 2 oz (50 g) almonds, blanched and shredded, a quarter of a grated nutmeg, and 1 teaspoon mixed spice. Moisten the whole mixture with 2 tablespoons sherry, or more if needed. Into all this fold 7 oz (200 g) flour, and place the cake in an 8-inch (20-cm) cake tin doubly lined with greaseproof paper. Place in a slow oven (300°F/150°C/mark 2) and bake for about 3 hours, testing when appropriate. When ready, leave for at least a day.

WHITBY YULE CAKE

Rub 12 oz (350 g) butter into 1½ lb (657 g) plain flour, and mix in ½ lb (225 g) moist brown sugar, ½ oz (15 g) cinnamon, a grated half nutmeg, ½ lb (225 g) raisins, ½ lb (225 g) currants, some candied lemon peel and as many almonds as desired. Beat up 3 eggs thoroughly with 4 tablespoons brandy, and mix this in with the other ingredients. Adding some cream, work the whole into a dough and press into a flat baking tin. Partly cut through the cake to form small squares, and bake in a moderate oven (325°F/165°C/mark 3) for 1½–2 hours. Turn out on to a wire tray and, when cold, break up where the knife cuts indicate. Serve each piece with a glass of cherry brandy.

Traditionally the cake eaten on Twelfth Night (6 January) Twelfth Cake has a French version in *galette des rois*, which is a less heavy cake:

TWELFTH CAKE

Cream ½ lb (225 g) butter with ½ lb (225 g) castor sugar and a little grated nutmeg. Beat in 4 eggs, one at a time. Stir in 5 tablespoons brandy. By degrees, sift in ½ lb (225 g) flour, and add 1 lb (450 g) well-washed currants, 3 oz (75 g) mixed candied peel, shredded, and 2 oz (50 g) blanched almonds. Line a large cake tin with greaseproof paper, place the mixture in it and bake in a slow oven (300°F/150°C/mark 2) for 3 hours. When ready, and cold, decorate with royal icing, made with 1 lb (450 g) sifted icing sugar into which a mixture of 2 egg whites and the juice of 1 lemon is slowly and thoroughly worked. When the whole is smooth and creamy spread a first thin layer over the cake with a knife dipped in cold water and let this dry. Then work a thick coating of icing on to the cake; with the remainder of the icing pipe decorations round the edges.

This is the French equivalent of pound cake, but with lighter measures:

QUATRE-QUARTS À L'ORANGE

Beat 3 large eggs with 6 oz (175 g) castor sugar until white and fluffy. Melt 6 oz (175 g) butter before folding it into the egg mixture when cool with 6 oz (175 g) sifted flour a little at a time, then 2 tablespoons orange curaçao and 3½ oz (90 g) ground almonds. Butter an 8-inch (20-cm) cake tin and cook the mixture for 30–40 minutes in a moderate oven (350°F/180°C/mark 4). Turn out and leave to cool. To prepare the icing, mix ½ lb (225 g) icing sugar with 1 tablespoon water, 2 tablespoons orange curaçao and the finely grated rind of 2 oranges. Split the cake in half, and sandwich together with some of the icing, mixed with ½ oz (15 g) ground almonds. Cover the top and sides of the cake with the rest of the icing.

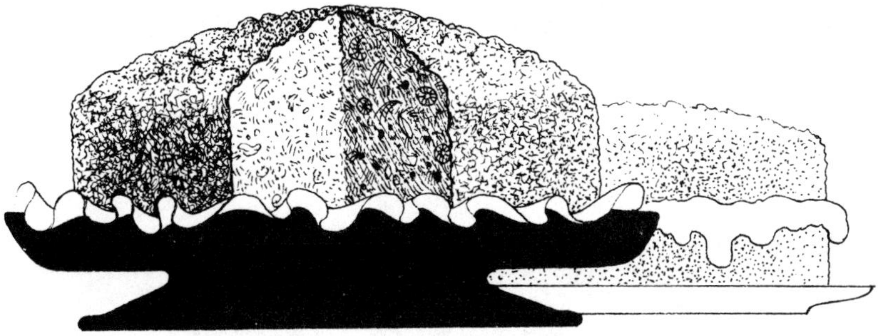

GÂTEAU ANANAS

Whisk together 4½ oz (115 g) castor sugar and 3 eggs over hot water, until very thick, and then remove the bowl from the heat. Continue whisking for a few minutes and then fold in 3 oz (75 g) plain flour, sifted, and 1½ oz (40 g) diced glacé pineapple, turning the whole mixture without delay into a large shallow cake tin. Cook in a moderate oven (350°F/180°C/mark 4) for 20–30 minutes. Let it cool, then split it horizontally. Cut a fresh pineapple into neat slices, sprinkle them with rum and castor sugar, and use most of the fruit to sandwich between the two cake halves. Make a hot apricot glaze from apricot jam and a little water, heated and sieved, and brush the outside of the cake with it. Decorate the top with the rest of the pineapple. Glaze again, thickly, and leave the glaze to set firmly.

Alsace is the home of this light French cake:

LE KOUGELHOPF

Use 3 tablespoons warm milk to dissolve ½ oz (15 g) dried yeast, and with 4 oz (100 g) flour make a soft dough, letting it rise to twice its original size. Place 10 oz (275 g) flour in a large bowl, making a central well, and put in this the risen dough, 3 oz (75 g) castor sugar, 3 eggs and a pinch of salt. Work these ingredients into an elastic dough, then work in 5 oz (150 g) butter and 4 oz (100 g) Malaga raisins. Now leave the dough in a ball in a warm place, floured, and covered with a cloth, so that it can rise. Butter a kougelhopf tin – a spirally fluted savarin tin – and sprinkle 2 oz (50 g) ground almonds round the inside. Half fill it with dough, cover, and let the dough rise to reach the top of the tin before cooking in a hot oven (400°F/200°C/mark 6) 40–45 minutes. Turn out and dust with icing sugar when cool.

Fruit Salads

MISCELLANIES of cold fruits, raw, chopped into pieces, soaking in their juice or in some fortified liquid, or resting on lettuce with selected vegetables and dressing, all require a marriage of inspiration and precedent, and an awareness of a few simple rules. The few types of salad are none too closely defined; a main group called *salades americaines*, which is distinctive in imitating the classic *salades vertes* but with the addition of fruit, mostly dating from the earlier part of the century and probably spurred on by the growing citrus fruit industry, is American in origin. The archetype is the Waldorf salad, devised at the great hotel, which has the following recipe:

WALDORF SALAD

Peel, core and slice 3 ripe eating apples. Take 1 lettuce heart, wash it and shred it, and slice 2 sticks of celery very finely. Season some thick mayonnaise with mustard and paprika, or red pepper. Sprinkle a little lemon juice and then some sugar on the apples, and arrange the salad on a serving dish. Add 2–3 oz (50–75 g) chopped walnuts, and chill the salad before serving. Serve the mayonnaise separately.

To this Boulestin adds (in a 1928 version) 3 chopped bananas, a diced half grapefruit or half orange.

There are at least seventy such salads in the standard *répertoire*; some idea of their composition is given by these summaries:

ALGONQUIN

Halved lettuces, dressed with grapefruit and pears, sprinkled with a julienne of green peppers, with a vinaigrette sauce mixed with chopped egg whites.

BEACH

Empty half a grapefruit, garnish the inside with lettuce leaves, replace grapefruit segments, decorate with cherries soaked in maraschino, and add a dressing of grapefruit juice, oil, salt and pepper.

GADSKI

Long, halved lettuces with sliced avocados, dressed with a dice of apples, green and red peppers, truffles and white grapes, with a vinaigrette sauce with crushed walnuts.

ORLÉANS

Grapefruit segments, with white and black grapes cut in halves and seeded, and pieces of pineapple, banana, orange and pear, all served on large lettuce leaves, surrounded with lemon slices; season with salt, pepper and cayenne pepper, and oil.

RIVIERA

Tangerine segments, chopped celeriac, chopped apples, strawberries and pineapple chunks, served on lettuce leaves with a mayonnaise with chopped sweet peppers.

SAINT-REGIS

Pineapple, grapefruit, celery, potatoes and asparagus heads, with a vinaigrette sauce with cream.

VENITIENNE

Olives, celery, diced truffles and orange segments; cook chicken livers in butter, sieve, and mix with green mayonnaise sauce to go with the salad.

There are a few matters to remember in making any fruit salad. All the fruit must be compatible and of consistent ripeness. A salad of fruits in its own juice, or with liqueurs, is much improved by marinading for 24 hours; however, if bananas are used, they must be added at the last moment. Lemon and lemon juice, and lime juice, are too pervasive to be used in salads, and on the whole berries and soft fruits are unwelcome. As appearance counts in the success of a fruit salad, dark-coloured fruit should never be included, looking particularly wrong in liquid salads. Fruit salads should be served in a china or silver bowl – never glass (reserved for trifles), which has the look of an aquarium.

Compotes are, broadly, fruits that have been cooked and have a syrup, served cold; macédoines are fruit salads where one or more fruits are cut into even-sized pieces – the name derives from Macedonia, presumably because it disintegrated into several republics in the second century. Francatelli used the description for vegetables, and it later became a common word for a medley of fruits in jelly.

Fruit salads belong to the later refinements of cookery. In Victorian times they are usually plain in concept but enriched with much cream and wine, sherry, Madeira, champagne or brandy. Indeed, to the following recipe, simplicity itself, Eliza Acton adds that 'persons who prefer brandy can substitute it for wine'.

PEACH SALAD

'Pare and slice half a dozen fine ripe peaches, arrange them in a dish, strew them with pounded sugar, and pour over them two or three glasses of champagne.'

PINEAPPLE FRUIT SALAD

One pineapple provides a fruit salad for 2 persons; it must be large, ripe and leafy. First cut it carefully lengthwise, right through the leaves. Then hollow out the flesh, leaving the shells undamaged and not pared so thinly from inside that any juice escapes. Wrap the shells in foil and chill until ready to serve. Dice the pineapple flesh and put into a basin with a mixture of 3 or 4 diced fruits – orange, pear, banana, pawpaw or strawberry, as available. Add a little sugar and a dash of lemon juice or brandy, and chill. When ready to serve, pile the fruit mixture into the shells, add a few whole strawberries on top, and serve with whipped cream.

SALADE SICILIENNE

Divide a number of tangerines into segments, removing as much of the fibre as possible from each segment. Take the same number of ripe pears and cut them into pieces matching the tangerine pieces in size. Sprinkle with castor sugar. Serve with cream.

The following is a splendid fruit salad devised by Colonel Kenney-Herbert, the author of a number of cookery books published at the end of the nineteenth century:

FRUIT SALAD AND CHAMPAGNE

Fill, but do not overcrowd, a handsome silver or old china bowl with a selection of perfectly ripe, choice fruit – 3 or 4 varieties if possible: large, carefully chosen strawberries, pineapple, bananas, peaches, apricots, etc. The pineapple pieces should be cut into fillets 2½ inches (6 cm) long and 1½ inches (4 cm) across. The peaches should be cut in quarters, the apricots in halves; the bananas should be slit in half lengthwise and then cut into 2-inch (5-cm) lengths. Arrange the fruit in the bowl and set this on ice. Make a syrup from sugar and water (using 1 lb (450 g) sugar to each 1 pint (600 ml) water), flavoured with old brandy and a little lemon, in sufficient quantity to lubricate the fruit, basting it with a few spoonfuls every now and then, turning the pieces gently. If kept over ice, the fruit will become encrusted with the syrup. For 15 minutes surround the bowl with ice and place ice on the cover of the bowl. Finally, pour in sufficient well-iced champagne to rise level with the top of the fruit, and send the bowl round.

DRIED FRUIT SALAD

A very large bowl is required. Wash the following fruits and place them in it: 1 lb (450 g) dried apricots, 4 oz (100 g) dried figs and the same amount of dried peaches, ½ lb (225 g) prunes, and 4 oz (100 g) raisins. Mix in 4 oz (100 g) blanched almonds, split in halves, and 2 oz (50 g) pistachios. Cover the whole mixture with water. Add

4 oz (100 g) sugar at least, or more to taste. A tablespoon each of rose water and orange-flower water can be sprinkled on. Now leave the fruit to macerate for 48 hours. Add some pomegranate seeds in their sacs to complete this Middle Eastern dish *(khoshaf)*.

FRUITS RAFRAICHIS

Boil together 1 lb (450 g) lump sugar and ½ pint (300 ml) water to make a syrup, dissolving the sugar completely before the mixture boils. Poach 1 lb (450 g) black cherries by cooking them for about 3 minutes in the boiling syrup, then place both cherries and syrup in a tureen. Cut up into small pieces a total of about 3 lb (1½ kg) different sorts of soft fruit – plums, greengages, peaches, pears or apricots, for example – and throw them into the tureen. Allow them to soak overnight, and before serving add a few drops of maraschino and some chips of ice. Have a jug of fresh cream ready to accompany the fruit, which should be sufficient for 12 people.

COMPOTE OF GRAPES

Peel 2 oranges as finely as possible. Make a syrup by boiling 20 lumps of sugar with 5 tablespoons water, dissolving the sugar completely before the mixture boils. When this is cold, add ½ pint (300 ml) champagne and the finely shredded orange peel. Now take 1 lb (450 g) white grapes, peel them, remove the pips, and pour the syrup over them. Chill the compote for 2 hours before serving.

COMPOTE OF ORANGES AND CHESTNUTS

Remove the rind, pith and outer membrane from 8 oranges, exposing the flesh. This is best done with a knife with a serrated edge. Cut into round slices. Arrange these in a glass dish so that they form the outer wall of a well, leaving the centre clear. Sprinkle with castor sugar and a little grated orange peel. Next peel 30 chestnuts (see p. 67) and cook them in water until tender; dissolve 1½ lb (675 g) sugar in ¾ pint (450 ml) water, boil until it pearls (at about 230°F (110°C), when a little syrup dropped in cold water forms a soft thread), then add a vanilla pod. Bruise the chestnuts, and cook in the syrup. Finally, place the chestnuts in the well, reduce the syrup until it is thick, and pour it over the compote.

COUPE GRIMALDI

Cut several slices of fresh pineapple into 1-inch (2½-cm) pieces, place them in a bowl and steep them in kirsch for 1 hour. Drain the fruit, place it in a glass, and cover with a tangerine *sorbet* (or alternatively an orange *sorbet* (see p. 138). Decorate with whipped, unsweetened cream and crystallized violets.

Alice B. Toklas's Indo-Chinese cook improvised this *coupe*. Desserts of this kind admit of unusual combinations, but their chances of perfect success – as opposed to being agreeably edible – are hazardous.

MACÉDOINE DE FRUITS (1)

Pare the rind of a small orange into ¼ pint (150 ml) water. Add 1 oz (25 g) sugar, dissolve sugar, then boil for 10 minutes. Add the juice of the orange, bring the syrup back to the boil, then leave it to get cold. Add 1 tablespoon sherry and set in a cold place. Next place in a bowl raspberries, strawberries, mulberries, grapes, peaches or apricots – as available – stoned and cut into equal-sized pieces. Pour the cold syrup over the fruit. Keep in a refrigerator until served.

MACÉDOINE DE FRUITS (2)

Take 4 ripe peaches and 4 ripe pears. Peel, and cut up into equal-sized pieces. Add a little lemon juice to prevent discoloration. Sweeten with sugar to taste. Decorate with redcurrants. Chill until very cold, and serve with cream.

Gooseberry

A PECULIARLY English fruit, the gooseberry is ignored by *la grande cuisine* with the one exception of a long established sauce for mackerel (see p. 246), which causes the French to give the gooseberry the name *groseille à maquereau* – *groseille* on its own being a red, white or black currant. *Ribes grossularia* is a native of Europe and North Africa not infrequently found growing wild, but domesticated by the time of Henry VIII, and almost fanatically encouraged in the mid-nineteenth century by the 'Gooseberry Clubs' of Cheshire and Lancashire, amateur growers who aimed to produce giant gooseberries. But these were always less well-flavoured than the smaller berries. For cooking, in fact, the small, green, acid fruits are the best; the great yellow and red berries are for dessert, best eaten fully ripe and warmed in the sun. A gooseberry wine made from the yellow berries is said to resemble champagne. There are white-skinned varieties, but these are not common; gooseberries may be dissimilar in shape and texture, with smooth, hairy or downy skins.

The first of the year's berries, the gooseberry traditionally provides the English Whitsunday fruit pie, and from the earliest times it appears as a sauce for meats, in dishes where its acidity counts, as in an Elizabethan recipe for chicken and gooseberry pie. Gerard also says that the berry is used in broths, a pleasant and wholesome addition. But its prime use must be in the hot gooseberry pie, a British contribution to culinary art, and in its counterpart, the chilled gooseberry fool (see p. 108).

OLDBURY TARTS

Put 2 lb (900 g) flour into a pastry bowl and form a well in the middle of it. Into this put ½ lb (225 g) butter and ½ lb (225 g) lard, both cut up into rough-shaped chunks. Have ¼ pint (150 ml) boiling water ready; pour it quickly over the butter and lard, stir with a knife until the fat dissolves, then stir in the flour to make a flexible paste. Flour a pastry board. For each tart to be made, set out two pieces of pastry, a large one and a smaller one. Roll out the larger to the size of a saucer, turn up the whole edge to about 1 inch (2½ cm) high, and cover the bottom with gooseberries. Over them put 1 large dessertspoon Demerara sugar. Roll out the smaller piece of pastry to about the size of a teacup rim, place it over the sugar and press the edges of the pastry firmly together to seal in the gooseberry syrup. Pinch round the edge to form fluting, and cook in a hot oven (400°F/200°C/mark 6) until crisp and browned.

GOOSEBERRY PIE

Moisten the edges of a large, well-buttered pie dish and place a strip of shortcrust pastry (see p. 11) round them; have the rest of the pastry, about ½ lb (225 g), rolled out ready. Take 2 lb (1 kg) gooseberries, young green ones that are small and sharp, and top and tail them. Place them in the pie dish, with sugar between each layer, using up to ½ lb (225 g) sugar, and piling them up into a mound above the level of the dish. Brush egg white on the pastry rim, and carefully lay on the pastry lid. Crimp the edges and brush the lid over with egg white. Dust with castor sugar. Bake for 10 minutes in a moderately hot oven (400°F/200°C mark 6), then reduce the heat to 350°F/180°C/mark 4 for 20 minutes. Serve the pie with ample thick cream.

BAKED GOOSEBERRY PUDDING

Top and tail 1½ lb (675 g) green gooseberries and stew them in a very little water. When they are tender, pass them through a sieve, adding ½ oz (15 g) butter, a beaten egg, 1½ oz (40 g) fresh breadcrumbs and sugar as necessary. Have ready a pie dish lined with shortcrust pastry (see p. 11) and fill it with the gooseberry mixture. Bake the pudding in a moderate oven (350°F/180°C/mark 4) for 40–45 minutes.

GOOSEBERRY CHEESE

Wash and dry a quantity of gooseberries that have been topped and tailed. They should not be fully ripe, but beginning to soften. Cook the berries in a steamer – without water – until they pulp, and rub them through a sieve. For every pound (450 g) of purée obtained, have 1 lb (450 g) sugar ready. Put the purée in a greased preserving pan, place over a low heat and stir until gently boiling. Boil for 2 minutes and then add the warmed sugar, stirring until it dissolves. Cook, stirring constantly, until the mixture thickens and no longer adheres to the sides of the pan. Now place in small pots and cover tightly; it will keep for up to 2 years and each pot can be turned out and eaten like cheese.

The name 'tansy' was given, in the seventeenth and eighteenth centuries, to the many fritters, pancakes and cakes which were flavoured by the shredded leaves of the hedgerow tansy, *Tanacetum vulgare*. Aromatic and bitter, the yellowish tansy had paschal significance, long surviving in an Easter tansy pudding made in Cheshire. Because of its bitterness, very few leaves were used, and in the course of time they were left out altogether; hence a number of dishes named tansy survived without the flower. Gooseberry tansy is noted in a recipe of Gervase Markham's in *The English Housewife*, similar to this later one of about 1700, but using sherry-sack and cream. The texture, which requires some care to achieve, should be like that of a firm omelette.

GOOSEBERRY TANSY

Cook $1\frac{1}{2}$ lb (675 g) gooseberries in butter in a covered pan until they turn quite soft. Beat 4 eggs, and stir into them 2 handfuls of breadcrumbs from a fine white loaf and ½ lb (225 g) sugar. Blend this into the gooseberry pulp over a slow heat. Stir gently. Do not let the mixture get too hot, and when it is firm turn out the tansy on to a hot plate. Sprinkle with castor sugar, and serve with cider or with some melted apple jelly.

NAMES: Eng. regional: carberry and wineberry (Yorkshire), feaberry (generally), groset (Scotland), etc.; Ger. *Stachelbeere*; Fr. *groseille (à maquereau)*; Sp. *grosella*.

STORAGE: Ripe gooseberries should be kept dry and cool and used within a day or two.

Grape

DESSERT grapes – *raisins de table* – belong to the species *Vitis vinifera* that produces wine grapes – *raisins de cuve* – and grapes used for drying, as currants, raisins and sultanas, and within this species there are varieties and complexities of classification without number. In differing forms, the vine has been cultivated since Noah planted a vineyard; thousands of years of wine-making have elevated the grape to a fruit of the highest economic importance, but its merits are not confined to wine. Grapes, as raw fruit, have nutritional and digestive virtues, with easily assimilable carbohydrates in the form of sucrose and fructose, with iron, potassium, sodium, calcium, magnesium and phosphorus, and with vitamin C and (in light-coloured grapes) B. The once famous grape cures point to their therapeutic value. But any gourmet would sacrifice this categorical knowledge for the sheer epicurean pleasure of eating a bunch of hothouse grapes, grown to perfection by management, better even than vineyard grapes still warm from the sun.

For dessert there are many choices, and mention need only be made of Spanish Almerias and Alicantes, of elongated Olivottes, of French Clairettes and Portugais bleu, of the Black Hamburgh, the

Frankenthal, the pink-coloured Rose Chasselas and the Chasselas de Fontainebleau, of Muscats and Frontignans. For cooking purposes Muscat grapes are usual, and Concord and Delaware in America are among the best of America's indigenous grapes.

Sun-dried or artificially dried grapes have a more widespread use in cookery, and the finer sultanas and raisins are eaten as dessert, often with nuts, in their own right. The three kinds are:

(1) *Currants.* Small, seedless, black, from the Corinth vine (Greek *apyrena*), from Greek Islands (Zante, etc.) and the Levant generally.

(2) *Sultanas.* Large, muscat-flavoured, yellow-brown, virtually without seeds, and retaining some sweet juice. The best are Turkish, once known as Smyrna raisins.

(3) *Raisins.* Malaga raisins, dried on their stalks, from Muscat grapes are best. 'Seedless raisins', in American nomenclature, are sultanas.

ALICANTE GRAPE SALAD

Take a good bunch of Alicante grapes, peel off the skins and remove the seeds; add a suitable quantity of walnuts, broken into quarters. Arrange the fruit on lettuce leaves and garnish the salad with maraschino cherries. Serve with French dressing.

SOLE VERONIQUE

Fold or roll the fillets of a Dover sole and season with salt and pepper. Place them in a buttered pan and almost cover them with a fish stock. Add 4 tablespoons curaçao. Poach the fillets very gently until they are done. Drain off the liquid in which they have been cooked, and with butter and flour make a sauce, suitably seasoned. Coat the fish with the sauce. Lay out the fillets on a serving dish, the edges of which have been decorated with vine leaves, and decorate each fillet with Muscat grapes, each one peeled and with the pips removed.

This is the Russian dish *kuropatka s apel'sinami i vinogradom*, suitable only for young birds shot in the same year they were bred:

PARTRIDGE WITH ORANGE AND GRAPES

Prepare a partridge in the usual way for frying, and fry in butter until brown. Place the bird in a deep dish. Chop the peeled rind of 1 orange finely, and blanch by pouring boiling water over it and draining immediately. Cut the orange itself into thin slices and cover the partridge with them, adding a handful of seedless grapes. Pour the butter from the frying pan over the bird, spread the orange rind over it, and bake in a moderate oven (350°F/180°C/mark 4) for about half an hour. When the bird is tender, serve it from the dish it has just been cooked in.

FAISAN AUX RAISINS

Have a pheasant prepared as for roasting, and put it in a deep pan in which some butter has been melted. Cook with the lid off, turning constantly until done. Then remove the bird and keep it hot. Meanwhile, take ½ lb (225 g) grapes, squeeze out the juice of half of them, and remove the pips and skins of the other half. Put the juice in the pan that was used for cooking the pheasant, season well, and scrape down any of the meat juice adhering to the sides of the pan; pour some of this over the pheasant, add 1 tablespoon warm brandy and set it alight. Next brown a mixture of half a finely chopped hard-boiled egg and some fine breadcrumbs in foaming butter, sprinkle it over the pheasant, and add the grapes around it. Pour in the rest of the gravy from the pan.

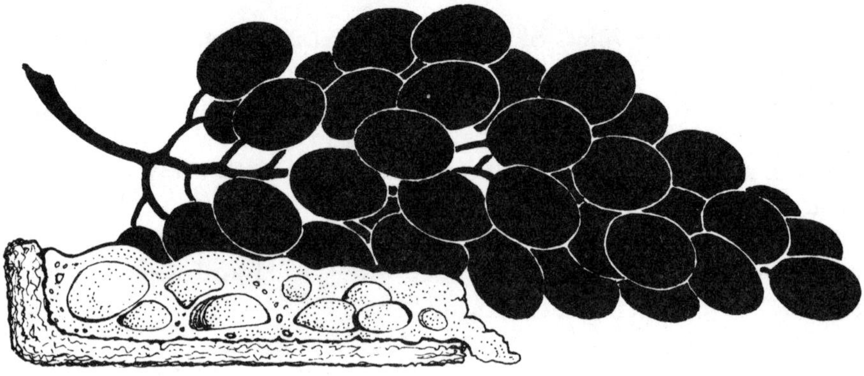

FOIE DE CANARD AUX RAISINS

Take a fresh duck's liver, whole and untouched, put it in a covered pan, season with salt and pepper, and cook it very slowly for half an hour with a little goose fat (goose dripping). Cook for a further 15 minutes with the lid off, after adding 4 tablespoons white wine. A few minutes before cooking finishes, add a handful of fresh grapes from which the skin and pips have been removed. Serve at once on toast. One of the excellent dishes of Gascon cooking, in which a goose's liver can be substituted.

TARTE AUX RAISINS

Sift ½ lb (225 g) plain flour into a bowl, and add a pinch of salt, 3 oz (75 g) castor sugar and 4 oz (100 g) butter cut into pieces. Mix these, and blend in an egg that has been beaten with a little water, until the pastry is smooth. Leave to chill. Have ready 1½ lb (675 g) grapes, washed, split and with the pips removed. These can be white grapes, or, if preferred, a mixture of white Muscat and black Alicante, or Chasselas. First place them in a dish to marinate in a few tablespoons of brandy and 3 oz (75 g) castor sugar for about an hour. Roll the pastry out to ⅛-inch (¼-cm) thickness, line a buttered 10-inch (25-cm) flan tin, prick the base and bake blind for 10 minutes in a moderately hot oven (400°F/200°C/mark 6). Meanwhile, mix 2 whole eggs, 1 egg yolk, 5 tablespoons milk, ¼ pint (150 ml) single cream and 2 oz (50 g) castor sugar together, with 1 tablespoon ground almonds and the juice drained from the dish of grapes: pour this mixture into the flan case when removed from the oven. Lastly, arrange the grapes in it, closely packed. Place in the oven, at a moderate heat (350°F/180°C/mark 4), for 20–30 minutes. The *tarte* can be served cold or warm, but not hot.

GRAPE PIE

From 10 oz (275 g) flour, 2 oz (50 g) butter, 1 teaspoon cream of tartar, ½ teaspoon bicarbonate of soda and a pinch of salt, make a paste, using milk to mix. Use half of it to line an 8-inch (20-cm) pie plate. Then wash 1 lb (450 g) grapes, cut them in half with a sharp knife and carefully remove the seeds, leaving the skins. Mix 2 tablespoons flour with 4 oz (100 g) sugar, sprinkle half of this over the grapes, then pour on 1 tablespoon lemon juice. Mix well, then sprinkle the rest of the flour and sugar on the bottom of the pastry case. Place the grapes in the case, dabbing butter here and there, and cover with the rest of the pastry. Bake in a moderate oven

(350°F/180°C/mark 4) for 30–40 minutes. The pie can be served hot or cold, with thin custard or cream.

GRAPE PUDDING

Cream 3½ oz (90 g) butter with 5½ oz (165 g) sugar and the yolks of 3 eggs. Mix in the grated rind of a lemon, a pinch of cinnamon and 2 oz (50 g) ground almonds, and fold in the crumb of 6 French rolls softened in warm milk. Wash 3 lb (1½ kg) grapes, slit them and remove the seeds, and add to the mixture, and finally fold in the whites of the 3 eggs, beaten until stiff. Put the whole mixture in a buttered pudding basin, cover tightly with greased greaseproof paper or foil, and steam for 45 minutes to 1 hour. Serve hot with a fruit sauce.

NAMES: Fr. *raisin*; Ger. *Traube*; It. and Sp. *uva*.
STORAGE: Grapes are ripe when picked, and should be eaten or used within a week.

Grapefruit

ALL citrus fruits originated in China or South-east Asia, and some were brought to Florida by the Spaniards in 1513, but one of the most important, the grapefruit (so named because it is borne in grape-like clusters), arose in the West Indies as a hybrid between the coarser pomelo and the sweet orange. The pomelo *(Citrus decumana)* was itself introduced to the West Indies by a Captain Shaddock (commander of an East India ship) in the seventeenth century, and it has been more often known as a shaddock. (The taxonomy is confusing. Some writers say the pomelo and the shaddock are the same fruit, some that they belong botanically to the same species but are horticulturally distinct.) Larger than the grapefruit, the pomelo has yellow or pink flesh, and sometimes dark red flesh, a thick rind that is easily peeled and segments that can readily be skinned. But even in the Bahamas it is a scarce fruit, as it does not attain the excellence of Thailand pomelos, grown on mounds irrigated with salt water, or other Eastern crops, and has largely been overtaken by the commercially grown grapefruit.

Although recognized as a species by about 1830, the grapefruit was still a novelty until the beginning of this century. By the 1920s this refreshing fruit, rich in vitamin C, was being consumed in huge quantities. A few connoisseurs of food stood out against a Chinese habit, copied by the Americans, of eating the fruit at the start of the meal, like P. Morton Shand, writing in 1927: 'I can no more be converted to beginning my lunch or dinner with half that parody of a lemon grafted on an anæmic orange ... than I can be induced to suck up cocktails through a straw.' But the fruit was entrenched and remains so, and its only drawback is that like all citrus fruits there is very little that can be done to it in the kitchen. Its juice can be extracted and is excellent; its segments can be conveniently tinned, and lose little value that way; it forms an agreeable part of many salads; marmalade can be made of it, the peel can be candied, and the fruit takes kindly to liqueurs. It is pleasant when warmed, but cannot actually be cooked.

It is perhaps most refreshing at breakfast. When Frank Hedges Butler was travelling in Cuba in 1920, he duly reported the following recipe, recommended as something new for English homes:

BREAKFAST GRAPEFRUIT

Cut a grapefruit in half, and treat it to a few drops of maraschino or sherry, and some lemon juice. Sugar as needed.

Here are three salads, out of many possible variants:

WEST INDIAN SALAD

Carefully cut up a grapefruit and divide into segments, removing the fruit from the membrane, and intersperse these with very thin round slices of banana and thin slices of tomato, peeled and with all the pips removed. On top of this goes a seasoning of finely chopped onion, and a cream dressing – cream, lemon juice, salt and pepper.

GRAPEFRUIT AND GRAPE SALAD

Halve and seed 1 lb (450 g) white grapes. Take the segments out of 3 halved grapefruit and 2 halved oranges, removing the fruit from the membrane, and combine all the fruit in a large bowl. In another bowl mix together 5 tablespoons lemon juice, 5 tablespoons plain yoghurt and 3 tablespoons clear honey until a consistent dressing results. Pour this over the fruit, and garnish the salad before serving with 5 tablespoons chopped walnuts.

GRAPEFRUIT AND STRAWBERRY SALAD

Peel and segment 2 pink grapefruit, removing the fruit from the membrane. Slice up 3 bananas, wash and hull $\frac{1}{2}$ lb (225 g) strawberries and halve them. Combine all the fruit with a cream salad dressing (see West Indian salad, above), and marinate in a refrigerator for 1 hour. Now drain the fruit and arrange on a bed of lettuce, garnishing with almonds and serving with a ginger dressing composed of $\frac{1}{4}$ pint (150 ml) sour cream, 2 tablespoons mayonnaise, $\frac{1}{4}$ teaspoon grated grapefruit rind and 1 tablespoon chopped root ginger.

BROILED GRAPEFRUIT

Halve a grapefruit and segment it, leaving each segment in the shell after detaching it from the membrane. Soften together sufficient butter and brown sugar to give each half a coating, and over this add a grating of nutmeg. Put the grapefruit under a preheated broiler or grill and leave it until it is hot and lightly browned.

GRAPEFRUIT PIE

Roll out sufficient flaky pastry (see p. oo) to fit over a deep pie dish. Allow an inch (2½-cm) margin all round. Cut another strip of pastry to fit round the edge of the dish. Dampen the edges of the pie dish and lightly press the strip of pastry round the dish, then brush it over with water. Combine grapefruit segments, pineapple pieces and orange segments sufficient to fill the pie dish, adding sugar as necessary. Fill the pie dish with fruit and cover with the rest of the pastry, without stretching it. Press the edges together with a fork, and trim off the surplus. Brush the top of the pastry over with milk. Bake in a hot oven (425°F/220°C/mark 7) for 10–15 minutes, reduce to a moderate heat (350°F/180°C/mark 4) and continue to bake for another half an hour. Serve hot.

GRAPEFRUIT AND PORT

Cut 2 grapefruit in half, and cut out the centres. Section the fruit by severing each segment from the membrane. Fill each centre with port, and chill thoroughly, for at least 4 hours. To prevent dehydration, each half must be wrapped tightly in a cellophane wrap.

NAMES: Grapefruit: Fr. *pamplemousse*; Ger. *Pampelmuse*; It. *pampelimosa; etc.*
Pomelo: derived from Dutch *pompelmoes.*

STORAGE: At room temperature or refrigerated. No further ripening takes place.

Guava

ONE of the commonest tropical fruits, the guava, an arborescent shrub *(Psidium guajava)*, is of the myrtle family and related to the clove. Tropical America is its home, and in Brazil and neighbouring countries it has been cultivated for a very long time, as well as running wild. It grows in parts of the United States, and has been known in the south of France and Algeria; an unexacting tree, the guava spread considerably through India when introduced from the West. There are a few varieties of guava, the best known being the Cattley guava *(P. Cattleyanum)*, with yellow or purplish fruits with white or reddish flesh and a flavour that has earned it the name of the 'strawberry guava'. This is the kind more generally preferred in the West Indies. The thick, yellowish-green skins of the ordinary egg-shaped guava reveal, when sliced open, a rich pink or salmon-red flesh, embedded with seeds.

It can be eaten raw, although there is a musky odour – which disappears on cooking – that some find a drawback. Its cookery uses are many in Central America. Lafcadio Hearn noted on his travels in the West Indies the delicious aroma of roasting parrots stuffed with guavas, but it is chiefly used for jellies, cheeses, pastes, jams, relishes and flavourings, as for ice cream or fools (see p. 109). The guava is rich in vitamin C.

This Mexican dessert consists of stuffed guavas:

GUAYABAS RELLENAS

Peel 8 ripe guavas and let them soak in water for about 1 hour, after which they should be drained and halved, and the seed centres and pulp scooped out. Put the shells in a pan, cover with water and cook for about 5 minutes. Set aside the guavas, now tender, in cold water. Now dissolve ½ lb (225 g) sugar in ½ pint (300 ml) water to make a syrup, bring to the boil, and add the pulp (if you are using tinned guavas and the pulp is not included, use an extra guava shell to make some pulp); simmer to reduce it to half. Take 3 medium apples, peel, core and quarter them, cover with water, boil until tender, and then drain. Add these to the guava syrup and cook very slowly until a paste is formed; use this paste to stuff the guava shells that have been set aside. Fit the 2 halves together to make a whole fruit, and sprinkle each one with finely chopped nuts. Chill before serving.

A South African Cape dish, this is sometimes known as 'Angel's Food':

GUAVA SALAD

Peel and slice 18 guavas thinly, arrange a layer of them in a glass dish and sprinkle a little sugar over them. Place a layer of thinly sliced oranges over the guavas, allowing 2 oranges for each layer, and again sprinkle with sugar. The next layer is guavas again, then oranges, and so on until the dish is full. Pour 4 tablespoons sherry over the fruit, and let the whole dish stand for a while before serving. Two tablespoons of orange liqueur may be added.

STEWED GUAVAS

Take as many guavas as are needed, and peel them as closely as possible. Cut them in halves, and scoop out the seeds. With water and a small amount of sugar, make a syrup in a pan. When it is cool, put the guava slices in it with care. Stew the fruit slowly, until it is soft, and serve hot, with cream.

GUAVA PIE

Prepare shortcrust pastry for a 9-inch (22½-cm) pie plate in the usual way (see p. 11), cover the plate, trim away the excess, crimp the edges and cook to a golden brown – 20–25 minutes in a hot oven (425°F/220°C/mark 7). Let it cool. Take 2 tins guava halves and drain the liquid into a saucepan. Boil and reduce the liquid by a third, and add to it 6 tablespoons lemon juice and 1 tablespoon arrowroot that has been dissolved in a little water. Cook this while stirring, until the liquid thickens, and then remove. Part of this glaze can then be spread over the base of the pie crust, and the guava pieces can be arranged in a circle on top, the rounded side upwards. Pour the rest of the glaze over the fruit. Let the pie cool. Serve with ½ pint (300 ml) whipped cream in a separate jug.

GUAVA MOULD PUDDING

Peel, slice and pulp some ripe guavas, add a little water and strain out the seeds. To every 1½ pints (900 ml) pulp obtained, add ½ lb (225 g) sugar; dissolve the sugar over a gentle heat, then let the syrup boil well for up to 10 minutes. Meanwhile, soak some gelatine, and add ½ oz (15 g) to every pint (600 ml) of the guava mixture. Heat the mixture gently if necessary to dissolve the gelatine, taking care not to boil, and pour into a mould. Let it set, and serve cold with cream.

NAMES: Aztec *xalxocotl* (= sand-plum); Sp. *guayaba*; Port. *goiaba*; Fr. *goyave*; Dutch *gujave*.

STORAGE: Ripe guavas will keep at most a day or two chilled.

Ices and Ice Desserts

REFRIGERATION, by means of ice pits and chambers, was well known in ancient China, and a kind of ice cream made from koumiss – mare's milk – prepared with flour and camphor and frozen was admired in the T'ang dynasty. The Romans may well have had some ice-cold desserts, perhaps snow flavoured with fruit juice. However, it is generally conceded that the first cream and water ices for sale were at the Café Procope in the Parisian street now called the rue de l'Ancienne Comédie in about the year 1660. The Sicilian Francisco Procopio, who engaged in this trade, was following in the wake of Catherine de' Medici's cook, Buontalenti, who had produced frozen confections for the Court over 100 years previously.

This essentially Italian art was furthered by Tortoni, whose establishment near the Palais Royal was famous in the late eighteenth century. Ices were now made all the year round; their flavour and refinement were much increased, and they had sufficient body to be moulded into shapes that are still standard. The *bombe glacé* is of this period, while under the Second Empire *coupes*, *mousses* and *parfaits* made their first appearance, as did the surprise *omelette à la norvegienne*, an ice cream encrusted with hot meringue which may have been invented by the scientist Count Rumford in 1804 during experiments in heat induction; later in the nineteenth century, Delmonico's in New York named the dish 'Alaska'.

In both America and England, Italian ice creams came with the emigrés, and the contemporaneous invention of the refrigerator in about 1870 in Sweden and the United States set the seal on commercial production of ice cream. Much was done domestically with freezers and ice-making gadgets, yet household refrigerators remained rare and expensive luxuries until the 1930s. By this time, manufactured ices had somewhat debased the possible delicacy and range of frozen desserts, the simple fruit-flavoured water ice, or *granita*, having almost disappeared. This was once served between courses at the more ample Victorian meals, and its refreshing character would not be out of place in modern menus.

There is a bewildering and extensive range of iced confections, but the basic knowledge of two forms, cream ices and water ices, will go a long way. Cream ices – invariably called ice creams – exist in two forms, with a mousse base or a custard base. In the first, dissolve 3 tablespoons of sugar in $\frac{1}{4}$ pint (150 ml) water to drawn thread stage, then add 3 egg yolks which have been beaten into 1 pint (600 ml) of cream. This is then churned in the freezer. For a custard base, whole eggs are beaten into a mixture of milk and cream, before being added to the sugar syrup as above. The basic water ice consists of 6 oz (175 g) sugar to 1 pint (600 ml) water, plus a fruit syrup for flavouring. Water ices (*sorbets* in France, so called from *sherbert*, a name sometimes used in America; *granitas* in Italy) are especially suitable for using fruit juices and flavourings derived from tropical fruits such as the mango, papaya, passionfruit or Chinese gooseberry. Indeed, among the many puddings, coupes and frozen mousses that can be contrived, there is hardly any fresh fruit or juice that seems out of place.

ORANGE GRANITA

Make a syrup of 3 tablespoons sugar in $\frac{1}{4}$ pint (150 ml) water, with 2 tablespoons grated orange rind and 2 tablespoons orange curaçao. Stir over a gentle heat until the sugar dissolves, then bring the liquid to the boil. Simmer for 5 minutes then cool. Put the syrup in a bowl, add $\frac{3}{4}$ pint (450 ml) fresh orange juice, and stir in 2 tablespoons curaçao until it all blends. Put the mixture in a freezer tray. Freeze until firm, then beat or blend the mixture until it is smooth. Freeze again for a few hours before serving.

LIME SORBET

Stir 12 oz (350 g) sugar into 1 pint (600 ml) water over a gentle heat until the sugar dissolves, then bring the syrup to the boil while continuing to stir, carefully dissolving any sugar crystals on the side of the pan. Simmer for 3 minutes. Grate enough lime rind to fill a tablespoon, add it to the syrup, and let it cool. Chill thoroughly for 2 hours, then stir in the juice of 10 or 12 sizeable limes. Strain the liquid into a bowl, freeze until firm, then beat or blend the mixture until it is smooth. Refreeze for a few hours before serving.

GOOSEBERRY SORBET

Top and tail and cook 2 lb (1 kg) ripe green gooseberries in 1½ pints (900 ml) water and 6 oz (175 g) castor sugar. Add the juice of 1 lemon and a little artificial green colouring, and pass the fruit through a sieve. First cool, and then freeze until the sorbet thickens. At this point add 4 tablespoons maraschino and 4 tablespoons white rum, beat the mixture until smooth, and then return to freezing.

TUTTI FRUTTI ICE

Dissolve ½ lb (225 g) white sugar in 1 pint (600 ml) water and boil for 10 minutes, then add ¼ lb (100 g) each of the pulp of ripe peaches, strawberry pulp and redcurrant pulp. Let this cool, then add 1 oz (25 g) very finely chopped candied peel and freeze the mixture.

ORANGE, BANANA AND LEMON ICE

Dissolve 1 lb (450 g) granulated sugar in 1 pint (600 ml) water and boil until a light syrup is obtained. Beat the yolks of 4 eggs lightly and add to the syrup, together with the juice of 3 oranges. Cook gently for 5 minutes, and set in the refrigerator to cool. Peel 1 banana, mash it with the juice of 2 lemons, and add to the custard. Then fold in 1 pint (600 ml) whipped cream, and without stirring replace in the refrigerator to freeze.

BLACKCURRANT LEAF ICE

Dissolve 1 lb (450 g) white sugar in ½ pint (300 ml) water, add 1 cup young blackcurrant leaves, and boil for 15 minutes, without stirring. Strain the hot syrup and pour it gently over the beaten yolk of 1 egg and the whites of 2 eggs. Beat until the mixture thickens, and as it does so fold in the juice of 1 lemon and ¼ pint (150 ml) whipped cream. Freeze and serve in glasses.

MRS MARSHALL'S APPLE AND SULTANA SORBET

Peel and slice 1½ lb (675 g) cooking apples, and place in a saucepan with 1 pint (600 ml) water and 6 oz (175 g) castor sugar, adding the peel of 1 lemon and 1 inch (2½ cm) of cinnamon stick. Cook until the apples are tender. Add the juice of 2 lemons and a few drops of green vegetable colouring. Pass through a sieve and allow to cool. Add 4 tablespoons brandy. Freeze to a thick batter-like consistency, then add ¼ pint (150 ml) thick whipped cream. Continue freezing, for 1½ hours. Meanwhile, prepare a compote of ½ lb (225 g) sultanas, a strip of lemon peel, a bay leaf, 3 oz (75 g) castor sugar and 4 tablespoons white rum, all simmered in ½ pint (300 ml) water. When the ingredients thicken and cream together, remove the bay leaf and the lemon rind, and chill the compote. Serve the sorbet with the compote spooned over it.

The *bombe*, so called from the original spherical shape of its mould (a conical one is now used), exists in well over 100 versions, and is now simply a lining (i.e. the outside) of one ice cream (or water ice) with an interior (the filling) of a light-flavoured *bombe* mixture. To make this mixture dissolve 1 lb (450 g) sugar in ¾ pint (450 ml) water, boil together, add 12 egg yolks and whisk over the heat until thick; sieve this, then whisk again until cold and pale.

Add an equal quantity of whipped cream, and the flavouring. Some permutations of the *bombe* are:

Bombe	*Outside*	*Inside*
Alhambra	vanilla	strawberry, base surrounded by kirsch-soaked strawberries
Americaine	strawberry	tangerine, decorated with pistachio nuts
Columbia	kirsch	pear, decorated with candied cherries
Comtesse Sarah	vanilla and kirsch	kümmel mousse, decorated with crystallized red and white rose leaves
Delicieuse	peach	*fine champagne*
Diable Rose	strawberry	kirsch and glacé cherries
Esperanza	orange	kirsch and *pralin*, with gooseberries
Frou-frou	vanilla	rum and candied fruits
Gladstone	ginger	gin-flavoured with dice of ginger and angelica
Jamaique	pineapple-rum	orange
Marechale	strawberry	three layers: pistachio, vanilla and orange
Medicis	peach flavoured with cognac	raspberry
Mireille	redcurrant	strawberry, kirsch and maraschino
Otero	apricot	blackcurrant
Sicilienne	lemon	vanilla-flavoured *pralin*, decorated with grilled almonds
Suzanne	rum, coloured pink	vanilla and Bar-le-Duc redcurrants
Trocadero	orange, dice of orange peel	layers of whipped cream and Genoese sponge soaked in curaçao, decorated with dice of orange peel

Of all the many *bombes*, Nesselrode is the best known. Count Nesselrode's ice pudding is said to have been invented by his chef, Monsieur Mouy, and to have excited the intense jealousy of Carême, Alexander I's chef.

BOMBE NESSELRODE

Peel 20 chestnuts and simmer them in a pan in 1½ pints syrup, made from 12 oz (350 g) sugar and ¾ pint (450 ml) water, with a stick of vanilla, until cooked. Drain and pass through a sieve. In another pan put 2 egg yolks, 2 oz (50 g) sugar and ½ pint (300 ml) boiled cream; stir and let the mixture thicken, but do not boil. Add the chestnut purée and 4 tablespoons maraschino. In another pan, cover 1 oz (25 g) stoned raisins and 1 oz (25 g) currants with syrup, cook them, then drain and cool them. Now partially freeze the chestnut purée then work with a wooden spatula. Add ¼ pint (150 ml) whipped cream, work again, then freeze completely. When the mixture is frozen, add the raisins and currants, put the pudding into an ice mould, and refrigerate for 2 hours. Now prepare a sauce from ¼ pint (150 ml) cream, with the yolks of 2 eggs and 1 oz (25 g) sugar. Stir these in a pan, until the mixture thickens but does not boil. Stir for 3 minutes after taking off the heat, then add 4 tablespoons maraschino. Chill the sauce and serve in a separate dish; the Nesselrode pudding is turned out on a napkin on a dish.

Another classic ice pudding is the following:

MONT BLANC

Roast and skin 1 lb (450 g) chestnuts, then stew them with 4 oz (100 g) sugar and 4 oz (100 g) chocolate and enough milk to moisten them. When they are soft enough, rub through a medium sieve into a mould to cool. Turn out when cold, and sprinkle with grated chocolate. Cover with whipped cream that has been sweetened and is fairly stiff: the chestnuts should be in a high mound wholly covered by cream, and the Mont Blanc served when thoroughly chilled.

Coupes – or sundaes (an American corruption of 'Sunday' occurring about 1904) – which are nothing more than composite dishes of ice cream and fruit, with perhaps a liqueur and nuts or crystallized fruit added, exist in many variations. The *coupe Jacques* is one of the most frequent to appear; of the many other possibilities, the following are a few established recipes:

AMBROSINE	Sliced peaches, soaked in kirsch, mashed with raspberry ice, surmounted by crème Chantilly (see recipe for *Nègre en chemise*, p. 18) and sprinkled with chopped roasted almonds.
ANDALOUSE	Segments of orange, soaked in maraschino, covered with lemon ice.
EDNA MAY	Vanilla ice and compote of cherries, masked with whipped cream flavoured with raspberries.
FAVORITE	A base of vanilla ice flavoured with kirsch and maraschino, bordered by pineapple ice, and in the middle whipped cream flavoured with strawberries.
NIÇOISE	Macédoine of fruits soaked in kirsch, covered by vanilla ice.
SEIN DE VENUS	Vanilla ice base, surmounted by a small poached peach, a strawberry set in the middle, the whole garnished with vanilla-flavoured cream.
VICTORIA	Macédoine of fruits soaked in brandy, covered with half a strawberry ice, half a pistachio ice.

COUPE JACQUES

In a champagne glass, or similar stemmed glass, or ice cup, place a macédoine of fruits soaked in kirsch, and on top of this half a strawberry ice and half a lemon ice. Between them place a peeled dessert grape.

Finally, one of the traditional English iced puddings:

LORD JOHN RUSSELL'S PUDDING

Take the yolks only of 6 eggs and beat them up with $1\frac{1}{2}$ pints (900 ml) milk, and place them in a pan with the grated peel of a lemon, $1\frac{1}{2}$ oz (40 g) sugar and $\frac{1}{2}$ oz (15 g) gelatine. Whisk over a slow flame, for about 20 minutes, taking care not to boil, until the gelatine is dissolved and the mixture is thick, then add $\frac{1}{2}$ pint (300 ml) single cream and 4 tablespoons brandy, and let it cook – but not boil – and thicken further. Then stir in a little crystallized pineapple, some candied citron and orange peel and a few crystallized or glacé cherries. Place the whole pudding in a 2-pint (1·2-litre) mould, cover firmly with greaseproof paper, and refrigerate. Serve this iced pudding very cold.

Jam

IT would be gratifying to believe with the eighteenth-century etymologist that jam took its name from nursery French, the child's '*J'aime*', I like it. Unfortunately it is simply a derivation from jamming, bruising or crushing, fruit, as was the habit with earlier recipes. Jam as we know it was a late phenomenon in British cookery, reaching its height at the end of the nineteenth century with formal afternoon tea, not simply in the Victorian home but in public (the first A.B.C. tea rooms opened in 1880). Jam's natural ally, bread and butter, had long featured on its own, in its own fashion – a Swiss traveller of 200 years ago complained that our bread was cut 'as thin as poppy leaves' – but it was not an English discovery: bread and butter was a combination copied from the Flemish. The commercial production of jam has done much to destroy the annual bout of domestic jam-making; it need hardly be said that homemade jam, properly made, is far superior.

Almost any fruit can be used to make a jam, some being traditionally more acceptable than others, and combinations of two compatible fruits enlarge the possibilities considerably. The pursuit of jam-making, however, is not the easiest of kitchen tasks, and there are certain rules, or at least maxims and precautions, to be heeded. Mrs C. F. Leyel, in *The Complete Jam Cupboard*, briefly summarized them in nine points:

1 The fruit and everything used in making the jam must be dry.
2 Always use a wooden spoon (a long-handled one).
3 The sugar should not be added until the fruit has been on the fire long enough to soften it, and then added only in small amounts at each boiling.
4 Jam should boil rapidly *before* the sugar is added, and quickly *after*, once the sugar is dissolved, to preserve colour and flavour.
5 All scum should be removed.
6 Jam is sufficiently cooked if it jellies when a little is placed on a saucer and allowed to cool.
7 Jam jars should be totally dry and clean, and should also be heated in the oven before receiving the jam.
8 The jam should be left a day after potting; a round of waxed paper should then be placed on top of the jam, and the cover tied on (or the screw-top put on). (Some advise putting the paper circle on the jam immediately it is potted).
9 All jams must be stored in a dry, cool place (the 'jam cupboard' of older households).

There are some other points to add:

10 A suitable preserving pan, of aluminium or enamel, and of large size, is a necessity.
11 Scum is best removed with a slotted wooden spoon.
12 Sugar to be used should be granulated, or, better than this, preserving sugar. Brown sugar should be confined, if used at all, to dark fruits.
13 A sugar thermometer is obviously a help.
14 Finally, large labels, with the date of making, the fruit content and all other relevant information, must be put on the jam jar.

Beyond this, more detailed information can easily be found about the finer points of preserving, including the subject of pectin. To summarize, the most readily made jams are those from fruits with a high proportion of pectin such as citrus fruit, gooseberries, blackcurrants and apples; a medium amount of pectin occurs in apricots, plums and raspberries, and a low pectin content in pears, grapes, pineapples, cherries and strawberries. This last group needs either the addition of some high pectin fruit in the making or the correct addition of commercially made pectin.

A plain apple jam is almost a rarity; too often, apples are used to fortify other fruits in jams, e.g. blackberries, improving these in flavour but not bringing out their own flavour. With so many varieties of apples available, the subtle possibilities of an unsophisticated apple jam should be better realized.

APPLE JAM

Take 4 lb (2 kg) russet apples, pare and core them, and cut them into small $\frac{1}{2}$-inch (1-cm) cubes or shapes. Weigh the cut-up fruit, and place it in a large bowl. Cover it with its own weight in granulated sugar and let it stand for a full 48 hours. Strain off the liquid that forms into a preserving pan and bring it to the boil. Now add the apple pieces, and slowly let them boil, without stirring, until they become transparent. Place in jars and cover when cold. A slight flavouring of 1 teaspoon grated ginger, or the juice and grated rind of 1 lemon to each pound (450 g) fruit, is permissible, but the apple will stand on its own merits.

BANANA AND PEAR JAM

Peel 6 lb (3 kg) bananas and cut them up into pieces of equal size. Peel and core 2 lb (1 kg) juicy pears, and cut them into pieces the same size as the bananas. Into a preserving pan put 1 lb (450 g) sugar, the juice of 2 lemons, and the pears; stir until the sugar is dissolved, bring to the boil, then add the bananas and another 1 lb (450 g) sugar by degrees. Stir carefully until the jam is boiling. Let it boil fast for 20 mintues, or until setting point is reached, skimming well. Pour immediately into pots and cover when cold.

CHERRY AND LOGANBERRY JAM

Put 2 lb (1 kg) black Morello cherries, stoned, with 2 lb (1 kg) carefully cleaned loganberries in a preserving pan, and bring gently to the boil, simmering slowly until the fruit is tender. Add 4 lb (2 kg) sugar. Stir it and let it dissolve, then boil, allowing 10–20 minutes until the jam will set. Pot it, and cover immediately.

CAMBRIDGE CHESTNUT JAM

Peel and skin 2 lb (1 kg) sweet chestnuts. Boil them, then pass through a wire sieve. Next make a syrup from 1½ lb (675 g) preserving sugar, ½ pint (300 ml) water and 2 tablespoons vanilla essence. Add the chestnuts to it and cook until fairly stiff. Pour into hot jars and cover when cold. Store in a dry place.

This recipe for *confiture de coco* was recorded from French West Indian sources by Stéphane Mallarmé; the poet published some Creole recipes in a women's magazine, *Dernière Mode*, in 1874 under a pseudonym.

COCONUT JAM

Take a copper saucepan and melt in it 1 lb (450 g) sugar with ¼ pint (150 ml) water. When the sugar reaches the brittle stage (at about 300°F (150°C)) add 4 oz (100 g) grated coconut, stir it in with a wooden spoon, and remove from the heat. A quarter of an hour later, put 2 egg yolks and a few drops of water in another saucepan and add the coconut mixture, constantly stirring in the same direction. Add a further 4 oz (100 g) grated coconut, flavour with vanilla, cinnamon or orange-flower water, and return to the flame for 5 minutes. Allow at least a further 5 minutes before serving, in a dish or pot, to be eaten with biscuits. Preferably the jam should be quite cold. This will not keep like a true jam.

DATE JAM

Stone 1 lb (450 g) fresh dates, chop them up into quite small pieces, and place them in ½ pint (300 ml) water. Let them simmer, stirring frequently, until tender. Add ½ lb (225 g) sugar and the grated rind of either a lemon or an orange and its juice, which should be strained. Dissolve the sugar, then boil until the jam is smooth and thick and will set. Pot, and cover when cold. Date jam does not keep well and should be used up fairly quickly.

FIG AND APPLE JAM WITH SPICE

Take 6 lb (3 kg) apples; peel and core them, and slice them up with 2 lb (1 kg) fresh figs. Place them in a preserving pan with the juice and grated rind of 3 lemons, ⅛ teaspoon each of grated or ground nutmeg, ground cinnamon and ground cloves, and ½ pint (300 ml) water. Cook all this very slowly, covering the pan. When the fruit is soft, add 6 lb (3 kg) sugar, let it dissolve, and bring the jam to the boil. Boil hard for about 15 minutes, and when it sets pot and cover without delay.

GREENGAGE JAM

Take greengages that are firm, but ripe. Wipe them clean, stalk them, and remove the stones. Weigh the remaining fruit, and set aside an equal weight of sugar. Take some of the greengage stones, crack them, and blanch the kernels. Add these to the fruit in a pan with not more than 1 inch (2½ cm) water. Gradually bring the fruit to the boil, and allow it to boil fast for 10 minutes. Put the sugar in. Stir thoroughly over a low heat until it dissolves, then boil fast for 20 minutes, or until the jam will set. Put in pots, cover when cold.

LIME AND GOOSEBERRY JAM

Top and tail 3 lb (1½ kg) gooseberries, and finely chop 12 limes. Put the fruit in a preserving pan with 4 pints (2·4 litres) water. Boil for 1 hour, and as the fruit softens mash it up. Add 7 lb (3½ kg) sugar and dissolve it over a low heat. Boil for a further 10 minutes, or until jam will set, then pot. As soon as the jam sets, cover it.

MANGO JAM

Take 4 fully ripe mangoes and peel them. Remove the seeds, and chop the fruit up completely, placing it in a preserving pan with 3 lb (1½ kg) sugar and the juice of 1 lemon. Dissolve the sugar over a low heat, then bring to the boil and allow 1 minute's hard boiling before removing from the heat. Stir in 5–6 fl oz (150–180 ml) liquid pectin without delay, then stir the jam and skim off the foam, and repeat this for several minutes. Ladle the jam into warm jars, and cover at once.

The small, golden yellow plum with its fine skin and pleasing flavour, the mirabelle, is more generally known in Germany and France than in England; a *confiture de mirabelles* from Alsace is an excellent jam:

MIRABELLE JAM

Take sound, ripe mirabelles and cut them in half. Extract the stones. Weigh the fruit, and for each pound (450 g) take ½ lb (225 g) sugar and ½ pint (300 ml) water and prepare a syrup from it. Place the fruit in the syrup, add a vanilla pod, and cook it, stirring all the time. As soon as it has boiled, reduce the heat and continue cooking slowly, for about 45 minutes. When the jam is the right consistency, pot and cover.

PEACH AND NECTARINE JAM

Take a number of very ripe peaches and the same number of ripe nectarines; stone them, skin them and then weigh the remaining fruit. Put all the fruit in a preserving pan with a very little water and cook quickly for 45 minutes. Care must be taken that it does not burn. Take it off the fire, and by degrees stir in 10 oz (275 g) sugar for every pound (450 g) fruit. For each pound (450 g) fruit, add also the strained juice of 1 lemon, and some of the kernels. Boil quickly for 5 minutes, stirring constantly. Remove the scum, pot, and cover when cold.

PEAR AND GINGER JAM

Peel 6 lb (3 kg) ripe pears, core them and cut them up into manageable pieces. Cover them with 4 lb (2 kg) sugar and leave overnight. Add 1 pint (600 ml) of apple juice the following day, and put in ½ lb (225 g) finely chopped preserved ginger. Put the pan over the heat and simmer the pears and ginger until they are clear and red. Pot the jam and cover when cold.

Raisines are simply jams which have grapes or grape juice reduced to a jelly-like consistency as a base. Apples, pears, melons, peaches or other similar fruit may be added, but they must be fruits of a kind to produce thin slices.

RAISINE DE BOURGOGNE

Take sound, ripe grapes and stem them; wash and seed them, and simmer, covered, with just enough water in the bottom of a preserving pan. Crush them through a sieve with a wooden spoon, gather the juice, and cook, skimming occasionally, until the *raisine* is reduced by about a third. Now take an equal quantity of ripe, sweet pears; peel and core them, and slice into fine slices. Add these to the *raisine* and simmer, stirring well, until the jam thickens. Add sugar to taste, pot, and cover when cold.

RASPBERRY JAM

Take ripe and dry raspberries, carefully picked from the stalks. Weigh them. Crush them in a bowl with a wooden spoon, for each pound (450 g) raspberries add ½ pint (300 ml) redcurrant juice – from currants previously cooked and strained – and place in a pan over a slow flame. Bring to the boil, then add the sugar, pound for pound (450 g) of the mixture. Stir until the sugar dissolves, boil up again, and in a few minutes the jam will thicken and set.

STRAWBERRY AND LEMON JAM

Select 3 lb (1½ kg) small sound strawberries. Place in a preserving pan and bring to the boil. Stir constantly. Warm, in the oven, 2¼ lb (1·1 kg) sugar so that it is ready to add to the fruit; add gradually and in small quantities so that the strawberries do not go off the boil. Pour in the juice of 1 lemon, strained. Boil together until the jam stiffens. Pot, and cover when cold.

ELIZA ACTON'S WHITE CURRANT JAM

Boil together quickly, for 7 minutes, 4 lb (2 kg) fine white currants, 'stalked with the greatest nicety', and 4 lb (2 kg) best sugar, pounded and passed through a sieve. Make sure the sugar is completely dissolved before the mixture boils. Stir the jam gently the whole time it is boiling, and from time to time skim it thoroughly. Test for a set. Remove and pot.

Jelly

FRUIT jellies – excluding the gelatine-based commercial 'nursery' jellies – lie somewhere between relishes or sauces and jams or desserts. They are made by a generally applicable principle of boiling the fruit, letting it strain through muslin, and cooking it with sugar until a clarified jelly is achieved. Jellies are suitable for preserving berries and fruits that are not in constant supply and are perhaps out of the ordinary, as they retain the essential flavour of the fruit. The most habitually used is redcurrant jelly (see p. 236), but other small fruit jellies are worth trying with game and other meats. To test for a set, put a little jelly on a cold saucer; if it wrinkles slightly when pushed with a fingertip, it is ready.

LOGANBERRY JELLY

Put 4 lb (2 kg) newly ripe loganberries and $\frac{1}{2}$ pint (300 ml) water in a heatproof jar. Cover closely. Stand the jar in a pan of simmering water until the fruit turns to pulp, and then strain, without pressing, through a jelly bag. For every pint (600 ml) of liquid so obtained, take 1 lb (450 g) sugar and 1 teaspoon lemon juice. Dissolve the sugar over a low heat, boil until the liquid starts to thicken, then pour into small jars. Cover the jars when the jelly has set and is cold.

MULBERRY JELLY

For each 1 lb (450 g) mulberries, carefully hulled, add 1 chopped up cooking apple, peeled and cored, and 8 tablespoons water in a pan, and bring slowly to the boil. Cook, after reducing the heat, until the fruit is soft, then remove the pan. Strain the fruit through a jelly bag, measure the juice, and for each pint (600 ml) add 1 lb (450 g) sugar. Heat the juice and dissolve the sugar, then increase the heat and boil for 10 minutes, when the jelly will reach setting point.

Lift off the scum, and pour the jelly into warmed jars. Cover when cold and store in a cool place.

This jelly can be served instead of redcurrant jelly with mutton:

SLOE AND CRAB JELLY

Put 4 lb (2 kg) ripe sloes in a pan with 4 lb (2 kg) crab-apples that have been quartered. All but cover with water, and boil to a pulp. Pour into a jelly bag and strain overnight. For each pint (600 ml) juice, allow 1 lb (450 g) sugar. Now bring the juice to the boil, boil for 10 minutes, and add the sugar. Stir until it is dissolved, then boil rapidly until jelly will set.

A Polish dessert, *Kieciel*, though not a preserve, is simple and adaptable to many fruit juices, such as redcurrant, raspberry, cherry and similar fruits:

KIECIEL

Strain about 1 pint (600 ml) fruit juice, and mix $1\frac{1}{2}$ oz (40 g) cornflour with, at first, a small amount of juice and then the rest of it until the mixture is wholly smooth. Add some castor sugar. Bring all this very slowly to the boil in a saucepan, stirring all the time. When it thickens, pour into a mould. Leave overnight to set. Chilled, *kieciel* should turn out like a jelly. Serve with sour cream.

Lemon

THE lemon, *Citrus limonium*, is the second most important citrus fruit after the orange. Introduced into the Middle East by the Arabs, the abundant lemon tree, fruiting more than once a year, is known to have been cultivated around Genoa in the mid-fifteenth century, and at the end of the century in the Azores, an early source for lemons imported into England. It is now widely grown round the Mediterranean, extensively in Italy and Spain, and in California. A sub-tropical fruit, which gives way to the lime in tropical conditions, the lemon has always been prized. In the age of Elizabeth I, John Gerard wrote that 'men in old time did not eat Citrons [lemons], but were contented with the smell, and to lay them among Cloathes, to preserve them from Moths'. Imported at high prices, aromatic lemons were too much of a luxury to consume, and so were sparingly used in the kitchen, the grated rind being added to one dish, the juice to another. Gradually they replaced vinegar for giving a sharper taste to food, and found their way in the eighteenth century into recipes for custards, creams, syllabubs, moulds and the like. The drink of lemonade, a French invention, was an innovation in England in Pepys's day. Nevertheless, though a few enthusiastic gardeners in noble households grew hothouse lemons, the fruit was to remain scarce until nineteenth-century shipping brought vast quantities into the country.

A bowl of Sicilian lemons belongs in every kitchen. Their predominant acid value enlivens a wide range of dishes, and the lemon is inseparable from Greek, Italian and North African cookery, from dishes of *hummus* to chickens stuffed with lemon. With most fish and seafood dishes, beyond the ordinary garnish of lemon slices, it is a courtesy to serve separately a few halves of lemon. The peel or zest is valuable in confectionery and puddings, and there are uses for preserved lemon peel, known as *citronnat* – the related citron *(C. medica)*, often referred to by its French name, *cédrat*, is almost exclusively used for candying and for crystallized peel. A sweet lemon, with the rind and flesh of a lemon, but with sweet pulp, is common in Italy, and there are departures from the unmistakable shape of the lemon in the over-developed 'Ponderosa' lemon of California and the 'Rough' lemon of the West Indies.

SIR FRANCIS ROSE'S LEMON SALAD

Cover some lemons with very salty water, boil until they turn soft, then drain off the water. When the lemons have cooled, cut them up. Mix with cooked artichoke hearts, cut up, salted almonds (see p. 15), 1 teaspoon honey and 2 tablespoons oil. Sprinkle lemon juice all over the salad.

AVGOLEMONO

This Greek lemon soup needs the addition of powdered saffron to give it a convincing yellow colour, but it basically consists of 2 pints (1·2 litres) chicken stock (or veal or rabbit). Heat the stock to boiling, then add the saffron, then the juice of 1½ lemons. Cool, stir in ¼ pint (150 ml) single cream, beaten with the yolks of 2 eggs, and add ½ teaspoon dry mustard. Stir well. Garnish with grated lemon peel. This soup can either be reheated – taking care not to boil – served hot, or chilled and served cold.

WELSH LEMON AND RAISIN PIE

Put 12 oz (350 g) raisins in a saucepan, and with them ½ lb (225 g) castor sugar, 1 oz (25 g) butter, 1 tablespoon flour, the juice and rind of 1 lemon and ½ pint (300 ml) water; cook them together until creamy, then let the mixture cool. Now line a plate with shortcrust pastry (see p. 11), spread the fruit thickly over it, cover with pastry and bake for 20–30 minutes in a moderately hot oven (400°F/200°C/mark 6).

LEMON PUDDING

Put ½ lb (225 g) butter, ½ lb (225 g) sugar and 3 tablespoons milk in a saucepan. Place over a gentle heat until the sugar dissolves, and when the mixture is cool add 6 well-beaten eggs and the juice of 2 lemons. Stir in the grated peel of the lemons and 2 tablespoons breadcrumbs, then pour into a pie dish lined with shortcrust pastry (see p. 11). Bake for 10 minutes in a moderately hot oven (400°F/200°C/mark 6) and then reduce the heat to 350°F/180°C/mark 4 for about 30 minutes or until set.

LEMON CHEESE PUDDING

Mix ½ oz (15 g) butter with 6 oz (175 g) sugar and 2 tablespoons flour, then add the well-beaten yolks of 2 eggs. Gradually add ¾ pint (450 ml) milk, and also, slowly, the juice of 2 lemons. Finally add the grated peel of 1 lemon. Mix all this well and fold in the stiffly beaten whites of the 2 eggs. Turn the mixture into a buttered pie dish, stand this in a larger dish of boiling water to prevent curdling. Bake in a very hot oven (475°F/250°C/mark 9) for 5 minutes, then reduce the heat and continue baking in a moderate oven (350°F/180°C/mark 4) for half an hour longer, or until set.

Here is a seventeenth-century recipe for what we would now call meringues:

LEMON PUFFS

Grate the rind of 2 lemons very fine and pound them in a mortar with 1 lb (450 g) castor sugar or mix in a blender; beat the whites of

3 eggs until stiff, mix them with the lemon and the sugar, and beat well together. Dust some sugar over some baking paper, drop on teaspoons of the mixture and bake in a cool oven (275°F/140°C/mark 1) for 45 minutes to 1 hour.

LEMON CURD

Melt 4 oz (100 g) butter, preferably over hot water in a double boiler, and stir into it 2 teaspoons grated lemon rind, up to $\frac{1}{4}$ pint (150 ml) fresh lemon juice, and 10 oz (275 g) granulated sugar. Continue stirring until the sugar dissolves. Now beat 4 whole eggs and 2 egg yolks, and blend gradually into the butter and lemon mixture. Cook until it is thick, then leave to cool and refrigerate.

NAMES: Arabic *laimūn*; Persian *līmūn*; Fr. *citron*; It. *limone*; Ger. *Zitrone*.

STORAGE: Choose lemons that are firm and have some green in the skin; they will keep and soften over a few weeks. Lemons of a bright consistent yellow may have been artificially ripened.

Lichee

A SLIGHT and beautiful tropical fruit (variously spelt lychee, litchi and lichee, from the Chinese *li-chi*), *Litchi chinensis* defied description by early travellers to the East. In 1824 Bishop Heber wrote of the fruit as 'a sort of plum, with the flavour of a Frontignac grape', and comparison with the better dessert grapes is predominant; however, Vaertijn in 1684 simply called lichees 'Chinese chestnuts', while a nineteenth-century English gardener speaks of 'the lichi hiding under a shell of ruddy brown its globes of translucent and delicately flavoured flesh'. Cultivation of this novelty to Western taste had in fact been undertaken in China for 2,000 years, and it had come to be the most luscious of fruits; the exiled poet Su Tung-po said that lichees, of which he ate 300 a day, would reconcile one to eternal banishment. The tree itself, a member of the Soapberry family *(Sapindaceae)* is long-lived; 100 years' profitable bearing is said to be in each.

The fruit grows in bunches, and the rind of each lichee, warty and a deep purplish pink in colour, conceals the pulp and a glossy seed; the pulp is white and edible, sweet and acidulous, traditionally the fitting last course of a meal in China. Apart from Asia, the lichee grows in Florida, in the Philippines, and sparsely in Central America, but it is mostly known to the world at large by the huge quantities canned for export from Asia. Fortunately, the tinned lichee preserves much of its taste, but the fresh fruit, seen occasionally in Western markets, is clearly preferable. It does not keep long, however. When dried for export, the pulp hardens and turns black, in which form the fruits are called lichee nuts; these are of no great interest, as they have lost their savour. A close relative, the rambutan, is much cherished in Malaya; it has the same general characteristics as the lichee, though on a larger scale, but it is not seen in the West.

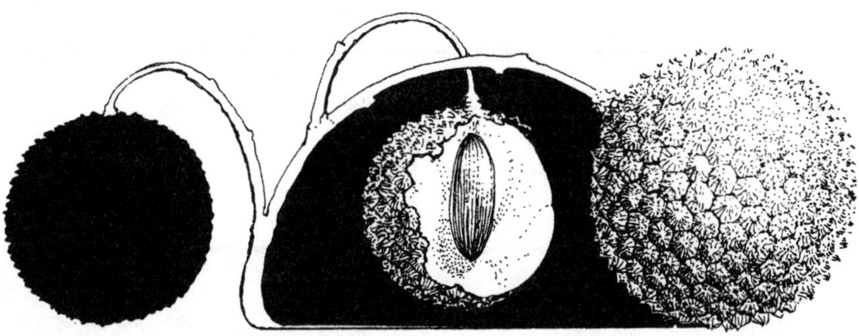

LICHEE AND ALMOND CHICKEN

Remove sufficient flesh from the bone of a cold chicken to fill 2 $\frac{1}{2}$-pint (300-ml) cups and dice it. Take 2 oz (50 g) almonds, blanch them (see p. 15), and split them when cool. Chop up finely 2 stalks celery, and slice a few spring onions as thinly as possible; put these with the almonds in a frying pan, with 1 tablespoon olive oil, and cook. When the almonds have browned, add the chicken and cook further for 5 minutes. Turn down the heat to low. Have ready $\frac{1}{2}$ pint (300 ml) lichee juice – obtained by expressing the contents of a whole tin of Chinese lichees and using the syrup – and mix half of it into the chicken, adding $\frac{1}{2}$ teaspoon ground ginger and some soy sauce, to taste. Turn the mixture with a wooden spoon, and fold in the rest of the lichee juice, into which 1 dessertspoon cornflour has been mixed. When the whole dish begins to thicken, serve hot with a mound of separately prepared rice.

FRESH LICHEE SALAD

Peel and remove the seeds from 48 fresh lichees. Lay tropical fruits in a bowl as a base: a freshly sliced mango, a chopped banana and a few thin pineapple rings would be suitable. Arrange the lichees on top. Prepare a light syrup, from 4 oz (100 g) castor sugar and $\frac{1}{4}$ pint (150 ml) water, flavour it with a little orange curaçao, pour it over the bowl, and let the fruit soak in it for an hour. Chill thoroughly, and sprinkle with kirsch. Serve with thin cream.

STORAGE: Keep cold, and serve almost immediately.

Lime

THE lime, *Citrus aurantifolia*, is the smallest tree of the citrus group and has the smallest fruits, green or yellowish green in colour. The West Indies have fruits larger than those that find their way to Europe, but they are thin-skinned and dry out quickly. But it is the juice that is the attraction, and this can be bottled. Lime juice, aromatic, sharp, and more persuasive than lemon juice, greatly improves fish and meat as a seasoning, enlivens soft fruits, and is inseparable from rum in the West Indies. Lime jelly is refreshing, and a distinctive marmalade can be made from limes. In general it can be put to all the uses of the lemon.

Some prominence attaches to the lime because of the naval surgeon James Lind's *Treatise on Scurvy*, published in 1754, which advocated its use at sea. It was not until 1795, a year after Lind's death, that Lord Hood added lime to the naval rations, some 200 years after the Dutch had learned the value of citrus fruits, and had actually introduced a mixture of lemon juice and the Mediterranean lime for use on sea voyages – the lemon being in fact richer in anti-scorbutic vitamin C than the lime.

LIME AND TREACLE TART

Line a flan case, or pie plate, with shortcrust pastry (see p. 11), and fill this open tart with a mixture composed of 1½ oz (40 g) breadcrumbs, grated finely, ½ lb (225 g) golden syrup, and 4 tablespoons freshly squeezed lime juice. Bake in a moderately hot oven (400°F/200°C/mark 6) for 20–25 minutes until the pastry is done and the mixture firm.

LIME SOUFFLÉ

Put 1½ oz (40 g) butter in a saucepan, heat it, and stir in 3 tablespoons flour. Let it cook for about 1 minute, then stir in ½ pint (300 ml) milk and 6 oz (175 g) sugar. Cook and stir for 3–4 minutes more. Have ready the strained juice of 3 fresh limes and their finely grated rind, and mix this in well. Let the mixture cool for a few minutes, then beat in the yolks of 4 eggs. Whisk the whites of the eggs separately with a pinch of salt until they stand in peaks, stir 2 tablespoons egg white into the mixture, and fold in the rest. Pour into a buttered mould of 2½-pint (1½-litre) capacity. Bake in a moderately hot oven (375°F/190°C/mark 5) for half an hour, or slightly longer if necessary. Serve at once.

NAMES: Fr. *limon*; Ger. *Limonelle, saure Limette*; It. *limone dolce*; Sp. *lima*.

STORAGE: Use within a few days, or keep in a refrigerator for slightly longer periods.

Liqueurs

USING liqueurs with cooked fruit is a question of matching the spirit to the fruit. Generally, but not necessarily, a fruit is complemented by the liqueur made from it or its kernels. When in doubt kirsch, the cherry-based liqueur, is advisable; as the table below indicates, it is more widely applicable than others, and it is available as *kirsch de cuisine* as well as in brands of better quality.

Fruit	*Liqueur*
apple	brandy, rum, kümmel
apricot	brandy, *crème de noyau*, curaçao, kirsch, rum
banana	rum, brandy, and most other liqueurs
blackcurrant	*crème de cassis*
cherry	cherry brandy, kirsch
fig	rum, kirsch
gooseberry	kirsch
grapefruit	maraschino
greengage	benedictine
orange	Grand Marnier, curaçao, rum, Van der Hum
peach	maraschino, kirsch, brandy
pear	rum, kirsch, brandy, maraschino
pineapple	rum, kirsch
plum	prunelle
prune	cherry brandy
raspberry	kirsch
strawberry	kirsch, cointreau, maraschino, Grand Marnier

Liqueurs are made from fruit or fruit kernels in one of two ways, either by distillation (as are all commercial liqueurs), or, more usually, by using a base spirit such as brandy, gin, vodka or tequila, which is as nearly 70 per cent proof as possible. This latter method produces what is in fact a ratafia, frequently encountered in cookery books from the seventeenth century onwards, and is the simpler way in the ordinary kitchen. Ratafias made of apricots, oranges, walnuts, quinces, grapes, cherries, lemons, tangerines, blackcurrants, raspberries, nectarines, peaches and plums are feasible on the lines of the basic recipe.

BLACKCURRANT LIQUEUR

At the bottom of a large jar place 1 clove, an inch (2½-cm) length of cinnamon stick and 12 small blackcurrant leaves, the greenest from the top of the blackcurrant bush. Wash 2 lb (1 kg) blackcurrants, drain off all the water, and put them in the jar. Crush them, pour over them 2 pints (1·2 litres) spirits (brandy or vodka) and mix. Add 1½ lb (675 g) white sugar and mix. Seal the jar and keep in a warm and preferably sunlit place for 1 month at least, and up to 45 days. Then strain the liqueur through a sieve into a bowl, expressing all the liquid from the fruit. Filter the liqueur into sterilized bottles and cork; drink the *cassis* after 2–3 months.

RATAFIA D'ORANGES

Grate the rind from 6 oranges and chop finely. Squeeze the juice into a wide-mouthed jar and add 1 lb (450 g) white lump sugar. Over this put the rind, a little crushed cinnamon stick and ½ teaspoon ground coriander. Pour 2 pints (1·2 litres) brandy over this, and stir. Seal well, leave the liqueur to infuse for 2 months, then strain and filter into sterilized bottles, leaving them sealed for at least another month before drinking.

CRANBERRY LIQUEUR

Take sufficient cranberries to provide 1 pint (600 ml) juice, and add to the juice the thinly peeled rind of an orange, an inch (2½-cm) length of cinnamon stick, bruised, and 6 cloves. Pour all this into a jar, add 2 pints (1·2 litres) whisky, cover tightly, and leave for a month to infuse. Then strain the liquid into bottles, and cork securely.

Loganberry

JAMES Harvey Logan (1841–1928), an American judge, produced this new fruit in about 1881, a hybrid with an undeniable character of its own. Judge Logan gave up the law in 1892 to devote himself to his experimental fruit and vegetable farm at Santa Cruz, and in due course presented the new bramble to the University of California; it took his name *(Rubus loganobaccus)*, and cultivation began. The loganberry reached Great Britain about 1900. It appears to have been a fortuitous crossing of the Wild Blackberry and the Antwerp Raspberry, although it is also claimed as a hybrid between the American Dewberry and the true Raspberry, and an upright-growing Dewberry (dewberry being properly the American name for the trailing rather than the bush blackberry); perhaps the loganberry is therefore a hybrid of *Rubus villosus* (dewberry) and *Rubus strigosus* (raspberry). Whatever its ancestry, this remarkably pleasant fruit has had a steady popularity as a summer berry to be eaten with cream, equal to that of the strawberry or raspberry, but on a more limited scale. A drawback of the fruit is its deep cavity, which is attractive to insects: every single berry must therefore be carefully picked over before use.

There are many less well-known hybrids, but they can mostly be treated as loganberries:

BOYSENBERRY	Possibly black raspberry and loganberry.
JAPANESE WINEBERRY	Small, sweet crimson berries; introduced from an oriental wild raspberry.
KING'S ACRE BERRY	Raspberry-like.
LAXTONBERRY	Resembles the raspberry.
LOWBERRY	Blackberry-like.
MAHDI BERRY	A raspberry-blackberry hybrid.
PHENOMENAL BERRY	Similar to the loganberry.
STRAWBERRY-RASPBERRY	Said to be 'devoid of taste'.
VEITCHBERRY	One of the best hybrids, a cross between a large blackberry and a raspberry.
WORCESTER BERRY	A cross between the gooseberry and the blackcurrant.
YOUNGBERRY	The dewberry crossed with the loganberry.

STEWED LOGANBERRIES

Dissolve 4 oz (100 g) granulated sugar in $\frac{1}{4}$ pint (150 ml) water and simmer in a saucepan. Add 1 lb (450 g) ripe loganberries. Cook gently for 5 minutes, allow to cool, and serve without breaking up the berries. Accompany with cream and sponge fingers.

LOGANBERRY SUMMER PUDDING

Take a number of thin slices of bread, a day or two old, trim the crusts off, and line a pudding basin with them; cut a circle to fit the bottom, and cut the other pieces into wedges to fit round the sides. Press together, leaving no gaps, and keeping enough bread to cover the pudding top. Put 4 oz (100 g) castor sugar and 2 tablespoons water in a pan, and dissolve the sugar over a gentle heat. Add $\frac{1}{2}$ lb (225 g) redcurrants and stew until tender, then add $\frac{1}{2}$ lb (225 g) raspberries and 12 oz (350 g) loganberries and stew for a few minutes more. Strain off some of the juice, about $\frac{1}{4}$ pint (150 ml), and turn the fruit carefully into the lined basin. Cover the top with bread, and then with a small plate, weighted, and leave the pudding to cool overnight. Turn out by loosening with a knife run round the sides of the basin. When serving pour the remaining juice over the pudding. Have clotted cream ready to go with it.

LOGANBERRY AND APPLE PIE

Take 1 large cooking apple, peel and core it, and grate it up coarsely. Mix the apple with 1 lb (450 g) loganberries. Put the fruit in a medium-sized pie dish with an egg cup in the middle, and cover it generously with 5–6 oz (150–175 g) soft brown sugar. Over this sprinkle the strained juice of 1 orange. Have $\frac{1}{2}$ lb (225 g) puff pastry ready (see p. 12), and roll it out to make a pie lid. Before placing in the oven, spread the white of an egg, unbeaten, over the pastry and sift castor sugar over that. Bake in a moderately hot oven (400°F/200°C/mark 6) for 10–15 minutes, when golden brown flecks should appear, then reduce the heat to 350°F/180°C/mark 4 and continue cooking until the crust is browned. Serve hot with cream.

NAMES: Fr. *ronce-framboise*; Ger. *Loganbeere*.
STORAGE: Loganberries rapidly turn mouldy if kept. Use within a day if fully ripe.

Mango

WHEN Akbar the Great planted an orchard of 100,000 mangoes near Darbhanga in northern India, he confirmed the value of this fruit to a people whose religion and mythology already accorded it an important role. Buddha himself reposed in a mango grove. But travellers to India were also enthusiastic over the royal fruit of *Mangifera indica*: John Fryer's nine-year travels, begun in 1672, led him to say of mangoes: 'When ripe, the Apples of the *Hesperides* are but Fables to them; for Taste, the Nectarine, Peach, and Apricot fall short.' There are, of course, degrees of perfection, and a poor, stringy mango bears the same relation to a fine, nurtured fruit as a crab-apple does to a melting Gravenstein.

The family to which this remarkable evergreen tree belongs is the *Anacardiaceae*; the mango has as relatives the cashew, the pistachio nut, the American poison ivy, the pepper-tree, and a few tropical fruits such as the imbu and the red and yellow mombins. There are about forty species of mango, several of which are cultivated for fruit throughout the world. The mango fruit is characteristically oval or kidney-shaped, with a smooth skin that varies greatly from green or yellow to orange or pinkish red; some have a crimson blush. The flesh is yellow or orange, with a large, flattened seed, some aroma, and a weight of anything from ½ lb (225 g) to 2 lb (1 kg).

Fresh, fully ripe mangoes (mangoes are ripe when the skin yields to slight pressure, regardless of colour) are probably best eaten with no addition; if any is needed, a little kirsch or cointreau will serve, with perhaps thick cream and a pinch of cinnamon. As the fruit is not easy to eat – the skin is stubborn if removed in the wrong direction, the flesh is juicy and slippery – it is best to serve the flesh completely removed from the stone and skin. It is difficult to describe the taste. It has often been compared to a combination of apricot and pineapple with, in unripe mangoes, a hint of turpentine. This disappears in cooking and in properly ripe fruit. It is the unripe, green mango that is pickled or preserved, and salted by the Indians as a chutney – the sweet mango chutney developed by the British in Viceregal India is not a feature of the native cookery. The ripe mango makes a succulent addition to fruit salads, and in purée form can be mixed with ice cream. It makes a good jam and a drink, and may be used in any recipe that requires a rich fruit pulp.

This green mango chutney, once an hors d'oeuvre of Martinique and Guadeloupe, ordinarily goes by the name of *rougail de mangues vertes*, a reminder of French colonial influence. It can be served with curry as a 'dip', or with fish or cold meat.

MANGO RELISH

Take 2 under-ripe green mangoes, peel them, remove the flesh, and chop it up coarsely. Put it through a sieve. Add 2 medium-sized onions, finely chopped, 1 small red pepper, seeded and chopped, salt and freshly ground pepper to taste, and a little olive oil to bind. The mixture can be further refined by pounding in a mortar; to heighten the taste add a little ground ginger and cayenne pepper. Chill the relish before serving, and garnish with chopped parsley.

MANGO MOUSSE

Take 3 large ripe mangoes, remove the flesh, blend it into a purée with the juice of 3 limes, and stir in 4 oz (100 g) castor sugar. Add 1 tablespoon gelatine that has first been dissolved in 2 tablespoons warm water and left to cool. Now beat 2 egg whites until they are stiff, fold in 5 tablespoons whipped double cream, and fold this mixture carefully into the mango mixture. Set the whole mousse in a serving dish and chill thoroughly for several hours before serving.

MANGO AND BANANA PIE

Slice 4 large mangoes and stew in a little water and sugar. Slice 2 bananas, and fill a pie dish with alternate layers of the stewed mango and banana. Sprinkle with sugar. Cover with shortcrust pastry (see p. 11) and bake for 20 minutes in a moderately hot oven (400°F/200°C/mark 6).

MANGO TART

Peel and slice 4 or 5 fresh mangoes, then place the fruit in a bowl with some chopped preserved ginger and 2 tablespoons ginger syrup and stir gently. Let this marinate for some time. With 4 oz (100 g) butter, 2 tablespoons castor sugar, 3 oz (75 g) flour and 3 oz (75 g) finely ground nuts, either hazelnuts or almonds, mix a smooth pastry, and line the bottom and sides of a 10-inch (25-cm) fluted flan tin. Bake blind in a hot oven (425°F/220°C/mark 7) about 12 minutes, until browned, and then cool. When ready to serve, line the tart with plain vanilla ice cream and cover the top with the marinated mangoes. Sprinkle with coarsely chopped nuts and serve very cold.

NAMES: Fr. *mangue*; Ger. *Mangopflaume*; It., Sp. *mango.*

STORAGE: At room temperature, the mango will soften and then may be kept only a few days more in a refrigerator; it should ideally be consumed as soon as ripe.

Marmalade

A MARMALADE is a preserve of boiled fruit and sugar that forms a consistent mass; it is often pointed out that the Portuguese for quince, *marmelo*, gave rise to another Portuguese word, *marmelada*, and so to the word that now generally means a preserve made of rinded fruit with pieces of the peel in it. There are early references to marmalades of quinces, dates, cherries, plums and, finally, oranges. What is marmalade to us is still referred to, in 1760, as 'preserved orange peel'. Tradition allows Scotland to be the home of modern marmalade, evolved rather than invented in Dundee by Janet Keiller, supposedly from acquiring part of a cargo of Spanish oranges too bitter to sell as such. The Keiller conserve reached Edinburgh and the recipe books; the Keiller preserves factory was established in 1797. Not until 1859 was finely shredded marmalade produced by a Paisley grocer named Robertson, and only in the 1870s did a Mrs Cooper of Oxford introduce this Scottish curiosity through her husband's grocery business to the University. It became popular at undergraduates' breakfasts as 'squish', and was eventually manufactured as Oxford marmalade, with its own distinctive character. Lastly, the firm of Chivers was founded in 1901, and the mass production of marmalade, enhanced by much annual marmalade-making in private households, was established in England.

It cannot be said that marmalade's great popularity has been echoed in other countries; it is hardly appreciated at all in America, and outside cosmopolitan hotels it remains an alien delicacy in France or Germany. This may be due to – to British minds – the evanescent nature of foreign breakfasts, their lack of toast, or of those traditional vehicles for orange marmalade, the buttered oatcake or the wheaten scone; again, it might gain a foothold in other cuisines if it were realized that marmalade, heated with a little white wine, makes a satisfactory sauce to serve with roast duck or goose, boiled or braised ham, or roast pork – and that, as a sauce, it can be fortified by the addition of a very little brandy or whisky.

There are a considerable number of ways of making marmalade, varying only in detail; oranges, lemons, limes and grapefruit are interchangeable in most of them. As with jam and jelly, test for a set by placing a small amount of marmalade on a very cold saucer. If it wrinkles slightly when pressed with a fingertip, it is set.

This first recipe is from *The Cook and Housewife's Manual*, published in Edinburgh in 1826, written by Mrs Johnston but issued under the name of Mistress Margaret Dods, the legendary president of the dining club at the Border Inn of which Sir Walter Scott wrote. It may be taken as the basic Scottish marmalade recipe:

MEG DODS'S ORANGE-CHIP MARMALADE

'Take equal weight of fine loaf sugar and Seville oranges. Wipe and grate the oranges, but not too much. (The outer grate boiled up with sugar will make an excellent conserve for rice, custard, or batter puddings.) Cut the oranges the cross way and squeeze out the juice through a small sieve. Scrape off the pulp from the inner skins and remove the seeds. Boil the skins [rind] till perfectly tender, changing the water to take off part of the bitter. When cool, scrape the coarse, white, thready part from the skins together for dispatch, cut them into narrow chips. Clarify the sugar, and put in the chips, pulp, and juice to it. Add, when boiled for ten minutes, the juice and grate of two lemons to every dozen of oranges. Skim and boil for twenty minutes: pot, and cover when cold.'

DARK THICK MARMALADE

Wash 12 Seville oranges and 1 lemon. Halve them and squeeze out the juice. Shred the remaining fruit thickly, and place the pips in a small muslin bag. Put all this into a large preserving pan with 1 gallon (4¾ litres) cold water; mark the height of the liquid and boil for about 2 hours, until the depth has been reduced by a little over one-third. Take out the bag of pips, add 8 lb (4 kg) sugar, stir until dissolved, bring to the boil, and continue boiling for a period between $1\frac{1}{2}$ and $1\frac{3}{4}$ hours. At this point the colour will darken and the marmalade will set.

CLEAR ORANGE MARMALADE

Peel the rind of Seville oranges very thin and cut it into strips. Remove the white, and cut up the pulp of the oranges. Throw the chips of rind into boiling water, boil them until they are tender, then drain them in a sieve. Pound them to a paste, and add this to the pulp. Steep the seeds in $\frac{1}{4}$ pint (150 ml) water, and add the jelly from this. Then boil it all together very quickly for 15 minutes without sugar. Next add $2\frac{3}{4}$ lb ($2\frac{1}{4}$ kg) sugar to every $1\frac{1}{2}$ lb (675 g) fruit, in 2 lots, boiling each lot for 15 minutes, stirring all the time.

TANGERINE MARMALADE

Take 1 lemon, 1 grapefruit and 3 lb ($1\frac{1}{2}$ kg) tangerines. Wipe them clean, and cut in half. Remove the juice and dispose of the centre core of the grapefruit. Then slice the rind of all the fruit thinly, and put about 10 oz (275 g) of the tangerine peel into a muslin bag. Put this, with the lemon and grapefruit rinds, and the lemon and tangerine juice and pulp, in a preserving pan, with 1 gallon ($4\frac{3}{4}$ litres) water and $\frac{1}{2}$ oz (15 g) tartaric acid. Bring to the boil, and boil very slowly until about half the contents of the pan have boiled away. With the excess of water driven off, strain through a linen cloth. The bag of shredded peel should have been removed after 45 minutes' boiling, and the shreds rinsed and dried in a cloth. Now put the strained liquid on the heat, stir in 3 lb ($1\frac{1}{2}$ kg) sugar until it has dissolved, bring rapidly to the boil, and continue to boil for 5–10 minutes, or until the marmalade sets when tested on a cold plate. Allow the marmalade to cool slightly, then stir in the tangerine shreds which have been reheated for a few minutes to sterilize. Put into pots when almost cold.

MRS BEETON'S LEMON MARMALADE

Place all the lemons to be used in a preserving pan, cover with cold water and boil gently for about 2 hours. During this time the water must be drained off and replaced with boiling water 3 times. Let the lemons cool slightly, and slice them thinly. Remove all the pips. Now weigh the fruit, and to each pound (450 g) allow 2 lb (900 g) sugar and 1 pint (600 ml) of the water the lemons were last boiled in. Dissolve the sugar in the water, and boil together until a thin syrup is obtained. Then add the prepared fruit, and boil until the marmalade jellies on a cold plate. Turn into pots, cover when cold, and store in a cool place.

LIME MARMALADE

Remove the rinds of 12 limes, and shred finely. Squeeze the juice into a pan with the rinds. Place the fine shredded pulp in a tied muslin bag and add to the pan, with 3 pints (1¾ litres) water. Boil gently for 1 hour. Then add 1½ lb (675 g) sugar, let it dissolve, bring back to the boil, and boil for 10 minutes or until it will set when tested. Pot the marmalade and cover without delay.

This marmalade was recorded in 1684, and was claimed to be 'as Madam Mancy made it for the Queen', meaning the late Queen Henrietta Maria:

CHERRY MARMALADE

Take 3 lb (1½ kg) stoned cherries, the juice of ½ lb (225 g) raspberries, the juice of ½ lb (225 g) redcurrants and 1 lb (450 g) fine white sugar, and place them together in a preserving pan. When the sugar has dissolved, boil quickly, and skim as the scum rises. Remove from the fire when the marmalade is 'of a fit consistency with a firm clear jelly mingled with the cherries'; bruise the cherries with the back of a wooden spoon, and when cold put the marmalade in pots.

PEACH AND PINEAPPLE MARMALADE

Pare and slice 1 large ripe pineapple. Peel and stone 7 lb (3½ kg) peaches, crack half the stones and remove the kernels. Put the peaches and the pineapple into a preserving pan with a little water, to protect the bottom, heat to simmering, then cook gently for about half an hour. Gradually add 6 lb (3 kg) sugar, without letting the temperature drop below simmering point, add the strained juice of 3 lemons and the peach kernels, and boil gently for about

20 minutes. Skim when necessary. Then pour into jars, cover when cold, and store in a cool place.

The kumquat is the least common of the citrus fruits, and is best used in preserves:

KUMQUAT MARMALADE

Take a quantity of kumquats, wash them, and slice finely, leaving the skin. Leave in cold water overnight, preferably in an enamel saucepan; the following day boil gently for 1 hour, and again leave overnight. Then measure the pulp and liquid, add $1\frac{1}{2}$ lb (675 g) sugar to each pint (600 ml) kumquat pulp and liquid. Dissolve the sugar by stirring over a gentle heat, then boil quickly until the mixture jellies when tested on a cold saucer. Cool a little, then pour into sterilized jars and seal.

MARMALADE TART

Line a flan case with shortcrust pastry (see p. 11) and spread on it a layer of marmalade. This can be orange, lime or pineapple. Next mix well together 3 oz (75 g) sugar, 3 oz (75 g) butter, 4 egg yolks and 3 whites, keeping 1 egg white aside, add 3 oz (75 g) flour. Spread this mixture on top of the marmalade. Bake for 10 minutes in a moderately hot oven (400°F/200°C/mark 6) then reduce the heat to (325°F/165°C/mark 3) for a further 30 minutes to cook the top. Now beat the remaining egg white until stiff and fold in 2 oz (50 g) sugar, spread this on top of the tart and return the tart to the oven for a few minutes. It will brown lightly.

This Scottish recipe makes a substantial pudding. A marmalade sauce made with brandy, and whipped cream, could be served in place of the custard:

MARMALADE PUDDING

Pour ¾ pint (450 ml) boiling milk over 6 oz (175 g) grated stale bread, and leave on one side until nearly cold. In the meantime, separate 3 eggs; beat up the yolks with 3 oz (75 g) sugar, add a generous tablespoon of marmalade, and stir into the bread and milk. Beat the egg whites into a stiff froth and fold them into the pudding. Pour the whole mixture into a buttered mould that has been ornamented with stoned raisins, cover with greaseproof paper or foil and place the mould at once in a pan of boiling water so that the water reaches the level of the pudding in the mould. Keeping the water at just under boiling point, cook the pudding for 1¾ hours. When ready, take it out of the water, lift off the lid, and let it stand for 5 minutes. Serve, after turning out, with hot custard.

Medlar

A STRANGE and attractive tree, *Mespilus germanica*, a member of the Roseaceae family, is sometimes found wild, with thorns, in southern hedgerows. It is usually planted for ornamental as well as fruit-bearing reasons. A native of northern Iran, now distributed throughout most of Europe, it was first known to the ancient Greeks. Hardy in growth, the tree is nevertheless so sensitive to the wind that it grows in deformed shapes; it has a brown fruit, the shape of a rose-hip and the size of a crab-apple, that is unusual in displaying five seed vessels and having its flesh set in a sort of leathery cup. When ripe, this can be eaten by scraping the flesh out of the skin – it is much enhanced by being mixed with cream and brown sugar – but 'ripe' here means half-rotten, or bletted, and

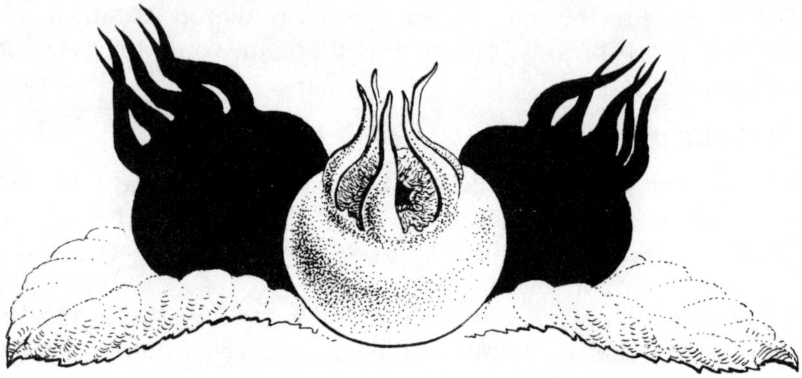

not before this; only when the white or greenish flesh has turned brown and is beginning to decay is the fruit soft and palatable. This means leaving the fruit to blet under the tree where it falls, or gathering it, brick-hard, and storing for several weeks; some Mediterranean countries, however, enjoy medlars that ripen on the tree.

The notion of decay dissuades some people from venturing on the fruit, while others find it uniquely agreeable; certainly in the late Middle Ages the medlar stood in high repute. It is said to be excellent with the best of wines. As the medlar is not commercially cultivated, the fruit must be found by whatever means possible. It is an autumn fruit, being picked after the first frost, and is sometimes marketed. After keeping, the fruit can be made into medlar jelly (see below) to accompany game or cold meats, or turned to various dessert uses – as medlar fool (see p. 109), or in any dish where fruit pulp can be utilized. Like a baked apple or roasted quince, the medlar can be set in a shallow dish with butter and cloves to bake for a few minutes before serving with thin cream.

LADY ADELA GOFF'S MEDLAR JELLY

When the medlars are ripe put them into a preserving pan with as much cold water as will cover them, and simmer slowly until they become a pulp. Strain through a jelly bag, and to each pint (600 ml) of liquid add 12 oz (350 g) sugar. Dissolve the sugar, then boil about 1 hour. When cold it will be stiff. Skim the top, but do not stir after it begins to boil. Boiling should be fast over a strong flame.

MEDLAR COMFITS

Place 4 lb (2 kg) ripe medlars in 1 pint (600 ml) water, and simmer until they are soft enough to be pressed through a sieve. Weigh the pulp after doing this, and add to it the same weight of granulated sugar. Stir the mixture over a low heat until the sugar dissolves, then bring to the boil, stirring with a wooden spoon, and let it thicken into a paste, thick enough to lift off the bottom of the pan as it is stirred. It is now cooked, and the dough should be placed on a pastry board sprinkled with castor sugar. Smooth the mixture with a knife into a flat, regular shape, and leave to dry for several hours. Then sprinkle with castor sugar, cut into comfit squares, and dip these in sugar before placing in paper cases and storing in a tin.

NAMES: Fr. *nèfle*; Ger. *Mispel*; Sp. *níspero.*
STORAGE: After picking, store in a dry cool place, on straw, calyx downwards; leave thus until bletted.

Melon

TRUE fruits, like the tomato, the melons all belong (except the water melon) to the *Cucurbitaceae* family, which includes cucumbers, marrows, pumpkins, squashes and gourds. These are considered vegetables. Dessert melons, of which many varieties exist, can conveniently be divided into 6 groups:

WATER MELON *(Citrullus vulgaris)* Native to Africa (Arabic *al-battikh* = Fr. *pastèque*), and the one differing species, it exists in many versions, the most familiar being the large green-black skinned melon with pink-red flesh and black seeds, the American 'green whale of summer', insatiably pleasing to eat when fully ripe, but not wholly satisfying. Like all melons it is virtually 95 per cent water; chilled wedges of water melon approximate to *granita* water ices.

MELON — *(Cucumis melo)* The melon is cultivated in most warm to tropical countries, sometimes under glass, frequently in the open, and exists in varieties of shape, colour, texture, thickness and pattern of rind, and colour and flavour of flesh. Charentais and Cavaillon melons are well thought of.

WINTER MELON — *(C. melo)* A smooth, yellowish-skinned, late-ripening melon, including the Honeydew melon with its pale green flesh. Their flavour is inferior to other melons, requiring ginger or some similar aid to enliven them.

MUSK OR NETTED MELON — *(C. melo var. reticulatus)* Round, with the familiar raised network; the flesh is aromatic. It is commonly a hothouse melon.

CANTALOUPE MELON — *(C. melo var. cantalupensis)* Said to have been raised at the Papal villa of Canteluppi near Rome, in the fifteenth century; warty skin, ridged, with orange aromatic flesh.

OGEN MELON — A recent export melon from Israel, named after the kibbutz where it was bred. Completely round, orange-yellow with green ribs and green flesh, it is small enough to be a desirable one-person melon.

The melon is not of ancient culture; only in the late fifteenth century did the fruit reach France, when Charles VIII brought it from Italy. Parkinson speaks of it as a rare and delicate fruit, eaten only by 'great personages'. These, indeed, consumed great numbers, in spite of its known indigestibility, which claimed the lives of two German emperors and Pope Paul II. Ideally, melons should be eaten completely ripe (difficult to judge with only one fruit, but tapping for a hollow sound is the surest guide) and *au naturel*, chilled. It is rarely cooked, but sometimes pickled (e.g. for a venison sauce). The attractive feature of melons is that they are their own bowl – a little hollowing out, the space filled with strawberries and kirsch, and there is an attractive *hors d'oeuvre* – and many different fillings, generally other fruit, can be imagined. Such *antipasti* as prawns in melon, or *prosciutto coi melone*, should also be remembered.

MELON SOUP

Take a medium-sized melon, a cantaloupe for preference, that is not yet quite ripe, remove the skin, and cut its flesh into small pieces. Cook these in butter for a few minutes, sprinkle some salt and pepper on them, and add a little chopped chervil. Add $1\frac{1}{2}$ pints (900 ml) boiling milk, bring back to the boil slowly, and let the soup simmer for 30 minutes. Now pass through a fine-meshed sieve, and bind with an egg yolk and some cream. Reheat, but do not boil. The soup may be garnished with *croûtons* of fried bread, cubes of melon fried in butter, and a few leaves of chervil.

CANTALOUPE AU MARASCHINO

Take a large cantaloupe, cut in half, and scoop out the flesh into a bowl, flavouring it with maraschino and castor sugar. When well iced, replace and serve in the rind.

CANTALOUPE AU PORTO

Make a hole in the top of a good cantaloupe melon and scoop out all the seeds. Sprinkle in a little castor sugar, and fill the inside with port. Place the melon on ice, and let it stand for 3–4 hours.

Pour out the port – keeping it for other purposes – and cut the melon into slices for serving. These should be as cold as possible.

MELON JELLY

Cut a Honeydew melon in half and remove the seeds. Make 1 pint (600 ml) jelly (using a packet jelly), adding 2 tablespoons rum, and chill it. Before it sets, remove from the refrigerator, whip up into a froth, and fold in $1\frac{1}{2}$ oz (40 g) grated coconut, a handful of raisins, and 12 glacé cherries, halved. Fill the melon halves with half set jelly, and leave them in the refrigerator to set and chill. To serve, hold the two halves together and cut the whole melon into slices.

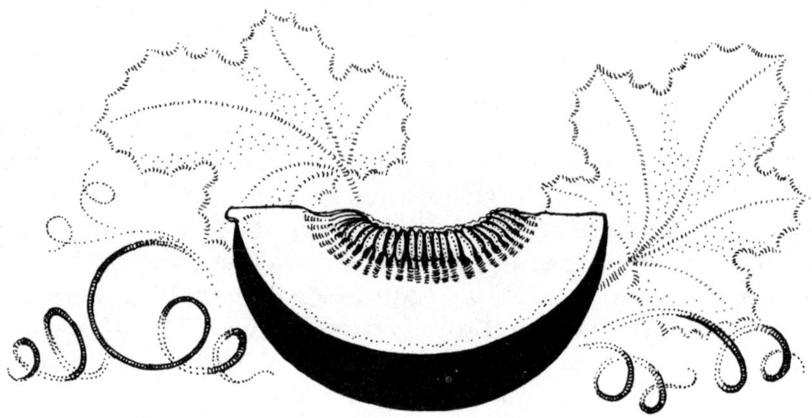

MELON EN SURPRISE

Cut a large cantaloupe melon in half, cut out the flesh and dice it. Add to it halved, stoned grapes, a dessert apple cut into small pieces, stoned cherries, sliced bananas or other fruit such as pine-apple, pear or apricot. Sprinkle with castor sugar and kirsch, replace fruit salad in the melon halves, and place these in bowls sur-rounded by crushed ice. Chill thoroughly before serving.

The next rather epicurean dessert derives from the manuscript recipe book of Alice B. Toklas's mother: the complex punch could be simplified by using only one liqueur:

MELON À LA SCHEHEREZADE

Slice off the stem end of a melon and scoop out the flesh, taking as much as possible while still leaving the melon intact. Empty out the juice, and cut the flesh into equal-sized pieces, also cutting a peach and 2 fresh pineapple rings into pieces of like size. Slice up a banana into very thin sections and add it to the fruit, together with a few small whole strawberries and some raspberries. Sprinkle with castor sugar, and when this has dissolved put the fruit into the melon. Mix the juice with a punch made of 4 parts dry champagne to 1 part each of kirsch, maraschino, *crème de menthe* and *sirop de grenadine*, and pour this over the fruit before the melon is placed in a refrigerator at least overnight. Serve the fruit from the melon.

NAMES: Fr. *melon*; Ger. *Melone*; It. *popóne*, *cocómero* (= water melon); Sp. *melón*.

STORAGE: In a cool, humid place for as long as it takes to ripen; then use at once.

Mixed Fruit

MORE than one fruit may go into a dish, and combinations of two fruits that complement each other are easily worked out. But some dishes involve many fruits (e.g. curries), and their success is not always assured; the flavours can merge and lose their identity, or simply conflict, or again take their flavour from one dominant fruit (lemon is to be avoided on this score). These recipes, which do not include cakes, suggest a number of ways of combining different fresh and dried fruits.

Mixed Fruit

This fruit curry forms the centre of a ring of cooked rice, and should be supported by small side dishes of, say, sliced hard-boiled eggs, lemon wedges, chopped ginger, or coconut:

FRUIT CURRY

Peel and slice 2 large cooking apples, 4 bananas and a large onion, and cut up 3 slices fresh pineapple, all into 1-inch ($2\frac{1}{2}$-cm) pieces. Seed 1 red pepper and cut into strips. Cook the onion in $\frac{1}{2}$ oz (15 g) butter in a pan, and a few minutes later add the apples, bananas, pineapple and red pepper. Cook further, then mix 1 tablespoon flour and 1 tablespoon curry powder to blend in the pan, and stir in 1 pint (600 ml) water until the mixture thickens. Add 6 stoned dates and 2 tablespoons sultanas, cover, and simmer for 10 minutes. The fruit should not break up. Add the juice of 1 lemon, 2 teaspoons sugar and salt and pepper to taste. Serve instantly surrounded by a border of rice.

This luxuriant dish comes from Brazil, duck with a fruit sauce:

PATO TROPICAL

Prepare a marinade for the duck, which should be jointed, consisting of one tumbler ($\frac{1}{2}$ pint (300 ml)) red wine, $\frac{1}{4}$ pint (150 ml) water, 2 tablespoons vinegar, chopped parsley, a sliced onion, 2 crushed bay leaves, a grated carrot, and salt and pepper; marinate the bird for about 12 hours. Then dry the duck pieces, roll in flour and sauté in butter in a saucepan. When lightly browned, add the strained marinade, together with the juice of 2 oranges, 4 stoned prunes, 4 stoned dates and 4 stoned dried apricots. Simmer until the duck and the fruit are cooked. Remove the pieces of duck to be kept hot in the oven, and pass the stock and the fruit through a sieve. Return this purée to the saucepan, adding a glass of sherry. Pour this sauce over the duck on an open dish; garnish with orange slices.

This old Sussex recipe, recorded by Philadelphia Shoebridge in her *Booke* of 1704, is more a fruit than a vegetable pie. 'Scaret' is what is now called skirret; once a popular and delicately flavoured root, it is hardly cultivated today. Substitute potato for skirret in this recipe. Preserved fruit probably means bottled fruit – bottled, fresh or frozen would do.

SCARET PYE

'First boil your roots and take the skin from them, season them with cinnamon, nutmegs, sugar and a little salt, not too sweet, the roots being so. Then butter the bottom of the pie, and lay on the potatoes or scarets, then lay on some preserved cherries, gooseberries, grapes, barberries, white bullace [you could use greengages] and [red or white] currants, some candied citron cut in thin slices, and lay over the other preserves marrow: two large [marrow] bones are enough for three pounds [1½ kg] of any of these roots. Then lay a pound [450 g] of good sweet butter over all and close the pie [with pastry]. An hour and a half will bake it.

'You must make the same caudle for this pie as you did for the artichoke, only squeeze in the juice of a good lemon before you put it in the pie.'

(This 'caudle' consisted of ½ pint (300 ml) white wine and ½ pint (300 ml) sherry boiled together, with a 'good piece' of butter added and the yolks of 3 eggs stirred in. This sauce is let into the pie when baked and shaken around before the pie is served.)

FLAN PANACHÉ

Make the flan case in the usual way (see p. 13), but with 2 eggs, tablespoon each of cream and water, the peel of half a lemon and a pinch of sugar, and bake for 15–20 minutes in a hot oven (425°F/220°C/mark 7). Then take 4 different kinds of fresh fruit – say, cherries, raspberries, gooseberries and blackcurrants – and poach ½ lb (225 g) of each for 4 minutes in a syrup made from 4 tablespoons water and 20 lumps of sugar. For very soft fruits, such as raspberries or strawberries, 1 minute's immersion when the syrup is lukewarm is enough. Arrange the fruits in quarters on the flan. Pass some redcurrant jelly through a hair sieve, add a few drops of brandy and cover the fruit with this glaze. Decorate with whipped cream and serve accompanied by crème Chantilly.

And finally an Austrian flummery of red fruits, 'red groats':

ROTE GRÜTZE

Take 1 lb (450 g) each of Morello cherries, redcurrants and raspberries, clean them, and cook with sugar until the fruit has softened sufficiently to be rubbed through a sieve. Reheat in the saucepan, mixing in 1 heaped tablespoon cornflour to which a little water has already been added, and stir for a few minutes. Leave to chill in a covered glass bowl. When cold, serve with whipped cream.

Mulberry

THE Common Mulberry, *Morus nigra*, is the mulberry that bears purple fruit; the Chinese White Mulberry *(M. alba)* is cultivated for its pale green leaves, food for silkworms, and not for its tasteless white fruit. There is a Red Mulberry *(M. rubra)*, native to America, a French Mulberry *(Callicarpa Americana)* with violet fruit, and in Japan the Paper Mulberry is cultivated for its bark. There are many old trees scattered throughout England, all, or nearly all, black fruiting trees, a benefice due to the followers of James I's edict of 1608 for the encouragement of silk production, who mistakenly planted *M. nigra*. Older examples of these large and very long-lived trees date from the first one planted in England, in 1548 at Syon House; then there was Milton's tree at Christ's College, Cambridge, and Shakespeare's tree – this was cut down in 1752 but Garrick grew a slip from it and one of its scions still grows at Hampton Court.

The fruit of the oldest trees is supposed to be the best; in any case, mulberries do not fruit for at least 15 years. The dark, purplish multiple fruit is possibly best eaten fresh from the tree with sugar and cream; it is decidedly juicy, difficult not to squash, and notorious for staining clothes and the person. For this reason, and because it has to be fully ripe when picked, mulberries are not grown commercially, and the old esteem in which the mulberry was once held as an ingredient of excellent jellies (see p. 151), jams, syrups, wine and shrub has all but disappeared.

This mulberry soup comes from Brazil:

SOPA DE AMORAS

Wash the fruit carefully and thoroughly, cover in white wine, and cook. Season with castor sugar and ground cinnamon. As the fruit disintegrates, serve hot over strips of toasted bread.

Mulberry

'A rede morreye', that is, a red mulberry dish, comes from Harleian Ms. 279 of about 1430 in the British Library in London. Wastil is bread of the finest flour:

A REDE MORREYE

'Take mulberries, and wring a good heap of them through a cloth; take veal, hew it and grind it small and cast thereto [i.e. mix together]; take good spicery and sugar, and cast thereto; take wastil bread and grate it, and yolks of eggs, and allay it up therewith [i.e. mix together], and cast good powder of spicery there-on above; and then serve it forth.'

MULBERRY FLAN

Prepare a flan case (see p. 13) and bake it blind. Stew 1 lb (450 g) clean, ripe mulberries in ¾ pint (450 ml) water with 6 oz (175 g) sugar. Allow to cool, and lift the mulberries into the case. Measure the syrup that remains, and dissolve gelatine in it in the ratio of ½ oz (15 g) gelatine to a pint (600 ml) liquid, adding 1 teaspoon lemon juice. Stir until cool. Strain over the fruit and leave while it stiffens. Finish by piping cream cheese over the flan.

MULBERRY PUDDING

Wash 1 lb (450 g) mulberries with care and hull them. Place them in a buttered pie dish with 4 oz (100 g) sugar, and grate over them the rind of a large lemon. Then add the juice of the lemon and the juice of 1 orange. Cream together 2 oz (50 g) butter and 2 oz (50 g) castor sugar, then work in 1 egg, 2 oz (50 g) flour, a few drops of vanilla essence and 1 tablespoon milk, and beat into a smooth mixture. Pour this gently over the mulberries and place in a moderate oven (350°F/180°C/mark 4) to bake for about half an hour. Serve hot, when the pudding is browned and cooked through, with cream.

NAMES: Fr. *mûre*; Ger. *Maulbeere*; Sp. *mora*.
STORAGE: Ripe mulberries should be used immediately.

Nectarine

FROM nectar, the drink of the gods, this accidental fruit, an unwitting but welcome sport of the peach, received its name in the seventeenth century, and it is so appropriate that it has been adopted by the French in preference to their own *brugnon* (from *prunus*). *Prunus persica* var. *nectarina* is the botanical name; the Italians know the fruit as *pesca noce*. Although separately cultivated, a nectarine sometimes occurs on a peach tree, and a peach on a nectarine. Its development in England owes much to the nurseryman Thomas Rivers. In such varieties as Lord Napier and Pineapple, it has become a dessert fruit considered by many to be superior to the peach. The main difference lies in the nectarine's lack of velvet; there is no *duvet* on the bright red skin. The white flesh, too, is firmer than the peach's, though perhaps with less texture – a nectarine is a fibrous, watery fruit, yet superior in flavour. Nectarines might seem too good to cook, but the sturdier flesh gives some advantage in this respect to a fruit to which all peach recipes can be applied; it is not a noticeably more expensive fruit and could be used more often for a more subtle taste.

NECTARINE AND ROCQUEFORT SALAD

Take $\frac{1}{4}$ pint (150 ml) mayonnaise and blend into it 4 oz (100 g) Rocquefort cheese, 3 oz (75 g) soft cream cheese, a dash of Worcestershire sauce, a pinch of salt and 1 tablespoon vinegar. Soften and dissolve $\frac{1}{4}$ oz (8 g) gelatine in a little water and stir this into the cheese, then put the mixture into about 8 small moulds, chilling them until firm. Arrange salad bowls with green salad bases, and in them place sliced nectarines – 4 or 5 large, fresh fruits – that have been dipped in an oil and vinegar dressing. Turn out the cheese moulds into the centre of the fruit.

NECTARINE STUFFED WITH CRAB

Take 3 or 4 ripe nectarines, and carefully sever the top to form a lid. Remove the stone with equal care, cutting it out with sufficient of the flesh to form a cavity. Into this stuff crab meat that has either been mixed with mayonnaise or rémoulade sauce, or chopped finely and mixed with chopped parsley, a very little chopped shallot, a chopped hard-boiled egg, mustard and some French dressing, or, finally, simply mixed with cream and seasoned with salt and paprika. However stuffed, replace the 'lid' and chill thoroughly before serving as a fish course.

NAMES: Fr. *nectarine (pêche à peau lisse)*; Ger. *Nectarine*; It. *pesca noce*; Sp. *nectarina*.

STORAGE: At room temperature, then chilled when ripe – although immediate use is advised.

Olive

THE ancient world rightly venerated one of the most important economic fruits of Europe – Homer mentions the olive groves of Greece. The olive tree *(Olea europaea)*, symbolic of peace and goodness, supplied the best of oils for sacred lamps and for medical, cosmetic and culinary purposes. The grey, gnarled tree with its green and silver leaves originated in the Mediterranean region, and has been cultivated there since Biblical times; in the time of Ptolemy II, as many as twenty-seven varieties are mentioned, and sophisticated farming methods and methods of extracting oil created a long-standing Mediterranean tradition. The olive grows, too, in other warm, temperate and arid regions, such as Australia, China and the Californian coast, producing a green fruit that turns to purple, the fruits mostly being picked when fully grown but still green. They are then steeped in water to reduce their bitterness, or pickled in brine. A dish of Spanish manzanilla olives with the stones removed, and stuffed with red pepper or a piece of anchovy, is a welcome balancing part of a Mediterranean meal, yet the fruit's appearance as such in cookery is only a token one (although traditional in such composite dishes as *paella*, and a good addition to game), its principal role being as refined cooking oil. That of Provence is said to be the best.

OLIVE AND FOIE GRAS TOASTS

Pound together 1 tablespoon butter and the same amount of *pâté de foie gras*, together with the yolk of 1 hard-boiled egg, and spread some of the mixture on rounds of toast. Take stoned green olives and stuff more of the mixture in them, place one on each piece of toast, and further decorate with chopped egg white.

COLONEL NOBLE'S OLIVE AND CAPER TOAST

Have ready 6 rounds of fried bread. Pass 6 anchovies through a sieve, mix them with a pat of butter and the yolk of a hard-boiled egg, and spread on the toasts. Chop finely 12 small green olives and

1 teaspoon capers small, sprinkle thickly on the toast, and place in a moderate oven (350°F/180°C/mark 4) for 10 minutes. Serve hot with a few drops of lemon juice on each.

OLIVE AND ORANGE SALAD

Peel and slice 4 ripe oranges. Stone 24 ripe green olives, and cut them up. Slice, into paper-thin slivers, a mild onion. Dress all these with a French dressing to which a dusting of cayenne pepper and a few rubbed basil leaves have been added. Garnish with a little fresh mint and some parsley. Chill before serving.

CHALOTAS CON ACEITUNAS

For a simple cold *hors d'oeuvre*, boil peeled shallots in a little water for 10 minutes – do not overcook – and serve garnished with green olives.

STUFFED OLIVES

Stone 6 large green olives carefully, and cut off one end to make them stand up. Fry 3 small rounds of bread – and stand the olives on them. Stuff them with a mixture made by pounding together, with a little butter and anchovy essence, 3 anchovies, 2 hard-boiled egg yolks, 1 teaspoon lime juice and a pinch of cayenne pepper. Season sharply, pipe a rose on each olive, and pipe the rest of the mixture on the bread. Chop the white of 1 hard-boiled egg finely and decorate the edge of the bread with it.

STUFFED OLIVES IN BATTER

Slit the end off and stone 12 large green olives, the largest possible (Spanish; those of Ascoli, noted for its *olive ripieni*, being virtually unobtainable), and set to one side. Take some cooked chicken – the white meat – and some smoked ham, both finely minced and mixed with a seasoning of salt and pepper, and brown lightly in butter. Sprinkle with grated nutmeg. When cool, mix in the beaten yolk of 1 egg and pack the filling into the olives, capping each with grated Parmesan cheese. Have sufficient fritter batter to cover all the olives (see recipe for Bundaberg banana fritters, p. 44); dip them in, and deep-fry in very hot olive oil. Serve very hot after draining on paper.

PLUVIER AUX OLIVES

Pluck and draw each plover (golden or grey) and put in each some green olives, themselves stuffed with red pepper. Truss the birds, brush lightly with fat, wrap bacon over each breast, and roast in butter in a moderate oven (350°F/180°C/mark 4) for 20 minutes, keeping the birds well basted. Serve on slices of bread fried in the butter used for roasting, with quarters of lemon, and, if needed, a brown sauce.

LAPIN DE BELLET AUX OLIVES

This dish was once a speciality of the country around Nice: wild rabbit, braised with tomatoes, tarragon, rosemary and green olives. The more usual *lapin aux olives* now consists of rabbit pieces browned in butter and oil, with onions and a garlic clove or two added. Add white wine, stock and flour already mixed with a little of the stock, tomato purée and mixed herbs, to make a sauce and to cook further. Add 4 oz (100 g) blanched olives halfway through the cooking – about an hour in all. Finally, add 4 tablespoons Madeira, and parsley to garnish. Serve the rabbit while still very hot.

NAMES: Fr. *olive*; Ger. *Olive*, (tree, *Ölbaum*); It. *uliva*; Sp. *oliva*; Gk *elaía*.

STORAGE: Use within a day or two before drying out begins; pickled olives will of course keep for very long periods.

Orange

THE most versatile and acceptable of citrus fruits, the orange has established itself in three forms at widespread intervals. In the eleventh century, the sour orange (*Citrus aurantium* and, earlier, *C. bigaradia*) was introduced into Italy and Spain from India. Four hundred years later, the sweet orange *(C. sinensis)* was brought, in one version, from Ceylon by the Portuguese and later, in a sweeter form, from China. This was the fruit that was cultivated in the orangeries of Europe – a fashionable gardening whim, as essence of orange was the modish perfume. Some ten miles from London, at Beddington Park, Sir Francis Carew grew the first orange trees in England, in about 1560, and over a hundred years later Pepys saw oranges growing in Hackney, still being nurtured in one of the many private gardens with a display of orange trees. Finally, in the late nineteenth century, the mandarin orange *(C. reticulata*, earlier *C. nobilis)*, came to England from China (see p. 258), but the essential picture of the orange in Western food is composed of the sweet and sour varieties. The sour orange (the name orange comes, through several mutations, from the Arabic *naranj*) was first known as the Bitter, Sour, Horned, or, commonly, Seville orange – this is the marmalade orange; the sweet orange as the Sweet, Coolie, Lisbon, Portugal, or, especially from the seventeenth to the early nineteenth centuries, China orange, and of these the varieties Jaffa, Valencia and Washington, and the Maltese and St Michael (both blood oranges), are well known among nearly a hundred sorts. Spain, North and South Africa, Israel and California are the principal areas of orange cultivation.

Apart from the immense quantity of oranges eaten out of hand, the fruit has always provided, from its zest, its juice or its segments, a distinctive contribution to cakes, puddings, pies, jellies, fruit salads, jams and sauces, the last being much used in the cooking of game, veal and some white meats. As a reliable provider of juice, and of healthy amounts of vitamin C, the orange is the source of one of the best of soft drinks, and two excellent liqueurs, orange curaçao and Van der Hum. Orange-flower water – the Italian Neroli is a distillation of this – is a flavouring of long standing.

This orange soup comes from Ecuador:

SOPA DE NARANJAS

Skin 12 oranges and scrape off the pith without breaking the flesh. Put in a large pan with 3 pints ($1\frac{3}{4}$ litres) water and 3 tablespoons sugar. Bring to boiling point slowly, turn down the heat and simmer for a further 20 minutes. Strain, and serve over toasted bread. Float very thin slices of lemon on the soup.

GRILLED MACKEREL WITH ORANGE AND OLIVES

Take 2 well-cleaned mackerel and cut slits on both sides of the fish. Lay them in a pan and pour over them a mixture of 4 tablespoons olive oil, 3 tablespoons white wine, and a little salt and freshly ground black pepper. Turn the fish in this preparation now and then, leaving it to marinate for about 2 hours. Then heat the grill, and place the fish under it for 5 minutes or more, until the mackerel are brown on both sides. Lift on to a dish, pour the pan juices over the fish, and garnish with split and stoned green olives, bay leaves and a whole orange sliced into rings. Serve with buttered brown bread and lemon halves.

CANARD À L'ORANGE

Clean the duck, sprinkle the inside with salt, truss, and sprinkle with salt and pepper. Have ready 2 oz (50 g) butter melted in a casserole; brown the bird in this, baste with up to 3 tablespoons warm brandy, and flame. Shred the thinly pared rind of 1 orange, and blanch it in boiling water. Let it cool. Set the juice of this orange on one side. Next prepare a caramel from 2 tablespoons sugar and 1 teaspoon vinegar; the sugar should dissolve slowly over a low heat. When ready, pour this over the bird, followed by the juice of half a lemon and the juice of the orange. Cover the casserole, return it to the oven, and finish cooking at a moderate heat (350°F/180°C/mark 4) for about an hour. While it is cooking, peel 2 oranges and slice them into rounds to place round the duck on the serving dish, shredding some of the rind as garnish, and adding any surplus juice to the duck sauce after cooking. Pour this over the bird before serving.

LADY SYSONBY'S BAKED ORANGES FOR DUCK

'With roast wild duck, serve whole oranges baked in their skins, and braised celery. Also rounds of hominy [a sort of cake made from maize meal], crisped on the outside with Parmesan and butter.'

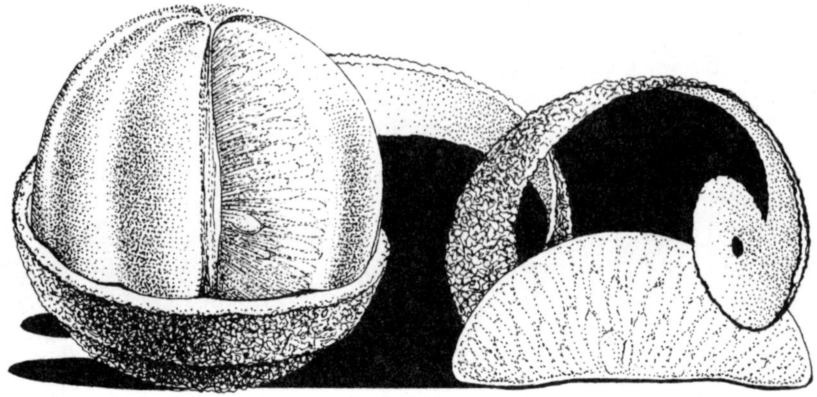

SHEPHEARD'S ORANGE SAUCE FOR DUCK

To the juices of simmering ducks, add 4 teaspoons grated bitter Seville orange peel, 3 tablespoons lime juice, a dash of Worcestershire sauce, 1 tablespoon curaçao, 6 tablespoons bitter orange marmalade, salt and cayenne pepper to taste, and 2 tablespoons brandy. Mix well.

Variations on orange sauces with duck or duckling are a matter of taste, but on the whole the more bitter sauces go better with the meat. Grapefruit peel can be substituted for the orange peel. Shepheard's, in its hey-day, garnished ducks with minute Jaffa oranges, halved, with the pulp mixed with honey, and the fruit dusted with brown sugar and grilled beside the birds.

BAKED STUFFED ORANGES

Take 4 sweet oranges and sever the stem end sufficiently far down to form a cap. Remove as much of the pulp as possible without damaging the skin. Chop up the pulp, and mix with 6 stoned dates and 4 dried figs. Blend in 1 tablespoon finely chopped almonds and the same amount of freshly grated white coconut, then stuff the mixture into the orange shells. Dust these with brown sugar, and bake in a moderate oven (350°F/180°C/mark 4) for half an hour. Add 1 tablespoon cointreau to each orange and splash a little heated brandy over the fruit. Serve flaming.

MRS MAYO'S ORANGE PUDDING

Butter a pie dish and line it with puff pastry (see p. 12). To prepare the filling, first boil the thinly pared rind of a large Seville orange in water until it is soft, then beat it in a mortar with the juice of the orange. Grate up 2 sponge fingers into fine crumbs, and add to the orange mixture. Cream ½ lb (225 g) butter and beat together with 4 oz (100 g) castor sugar, then add the yolks of 6 eggs. Beat the whole mixture together and pour into the puff pastry. Bake for 20–30 minutes in a hot oven (425°F/220°C/mark 7).

CRÊPES SUZETTE

Make a thin batter, beating 3 eggs with 6 tablespoons flour and then adding ½ pint (300 ml) milk. Cook the crêpes in melted butter in a small frying pan, keeping the crêpes very thin. Set them aside and keep warm. For the sauce, place in a shallow heated pan 2 oz (50 g) butter, 6 oz (175 g) sugar and 2 tablespoons kirsch, and heat slowly to dissolve the sugar. Squeeze in the juice of 2 oranges, and add the peel of 1 lemon and 1 orange, finely grated. Mix in, and let the sauce simmer until it becomes syrupy. Lastly, add 1½ tablespoons cointreau and the same amount of Grand Marnier. Put the crêpes in the pan, one by one. Cover each crêpe on both sides with the sauce, fold in half and then in half again. Pour on 1 or 2 tablespoons brandy. Agitate the pan, tilting back and forward, until the brandy first ignites and then is extinguished. Serve. Pour the remaining sauce from the pan over the crêpes.

ORANGE CARAMEL

Cut the rind, pith and outer membrane from 6 oranges, using a knife with a serrated edge. Put in a bowl as for fruit salad. Prepare a caramel by dissolving 20 lumps of sugar in 2 tablespoons water, and heating it until it is dark brown. Then spread the caramel over a large buttered plate; when cold, just before serving, break it up as finely as possible and sprinkle it over the oranges. Whip up ½ pint (300 ml) double cream stiffly, cover the oranges and caramel with it, and lastly sprinkle 2 oz (50 g) coarsely chopped blanched almonds over the dish.

ORANGE JUMBLES

Cream 5 oz (125 g) butter with 5 oz (125 g) castor sugar. Mix in half a beaten egg and add 5 oz (125 g) sifted self-raising flour, 1 teaspoon grated orange rind and 1 oz (25 g) ground almonds. Form into rolls ½ inch (1·5 cm) in diameter, cut these into pieces 4 inches (10 cm) long and form into S shapes. Bake in a moderate oven (400°F/200°C/mark 6) until pale golden brown and allow to cool a little before transferring to a wire rack.

NAMES: Fr. *orange*; It. *arancia*; Ger. *Apfelsine*, *Orange*; Sp. *naranja*.
STORAGE: Ripe oranges keep for weeks; when mould attacks them, it is at once evident.

Papaya

THE papaya, also familiarly called the pawpaw, is the fruit of the tropical American (Brazilian) plant *Carica papaya*, not to be confused with the papaw (or pawpaw) which is also an American food-plant and a member of the custard-apple genus *(Anonaceae)*. The real papaya, apart from its native continent, grows in India, Malaya, Tahiti, Australia, South Africa, Haiti, Mauritius and similar areas. A tree-like herbaceous plant with leaves like those of the fig, the papaya produces a very large berry, the size of a melon, oblong or round, smooth or ribbed on the exterior. The flesh is also like that of the melon in taste (it has been described as having a peculiar cowslip-like flavour) and in its pink to orange colour and its mass of bunched black seeds. Turning from green to orange-yellow, the papaya becomes sweet and juicy, an excellent dessert fruit with valuable digestive qualities, its pulp being a source of papain, the enzyme valuable in breaking down protein. The leaves of the papaya also act, when wrapped round it, as a meat tenderizer. The green fruit is treated rather like a marrow, and can be boiled or pickled; the ripe fruit is eaten with lime juice, or with ginger and sugar, and also lends itself to a number of vegetable-fruit uses. A prolific and easily grown fruit, the papaya is well established as one of the more important tropical and sub-tropical fruits.

PAPAYA AND PINEAPPLE SALAD

As a dressing, blend 1 tablespoon of either lemon or pineapple juice with $\frac{1}{4}$ pint (150 ml) mayonnaise and chill this for an hour. Peel and seed 1 ripe papaya and dice it, and dice the flesh of a large fresh pineapple. Mix the fruits, and pile into a salad bowl lined with fresh lettuce. Add some of the dressing, and serve the rest separately.

STUFFED PAPAYA

Take 1 medium-sized green papaya, wash it with care, then cut off the end and scoop out the centre and the seeds. To prepare the stuffing, chop up 2 small onions, 2 tomatoes, 1 sweet green pepper, 1 small hot red pepper (chilli), and 2 cloves of garlic, and cook them in butter for about 20 minutes, until soft. Grill 2 rashers of bacon, chop these up finely, and add to the vegetables. Take 1 tablespoon breadcrumbs, mix with 1 beaten egg, some seasoning and the vegetable and bacon mixture, and stuff all of it into the papaya. Skewer on the end with wooden toothpicks, and bake the whole fruit in a covered dish, with $\frac{1}{4}$ pint (150 ml) water, for about 45 minutes at a moderate heat (350°F/180°C/mark 4). Serve hot; the outer papaya skin should be left uneaten.

PAPAYA AU GRATIN

Peel and cut into sections a green papaya, and boil it in salt water until tender. Mash thoroughly into a pulp. With 1 heaped tablespoon flour, 1 oz (25 g) butter and $\frac{1}{2}$ pint (300 ml) milk, make a white sauce and add to it 4 oz (100 g) grated cheese; mix all together, and pour into a well-buttered baking dish. Cover well with breadcrumbs, and cook in a moderate oven (350°F/180°C/mark 4) for 20 minutes.

PAWPAW PUDDING

Peel a small papaya and cut up into 1-inch ($2\frac{1}{2}$-cm) cubes, and boil them in water until soft. Add sugar while they are cooking, and cinnamon to taste. When sufficiently soft, drain the papaya, place in a pie dish, and over it pour an egg beaten with $\frac{1}{4}$ pint (150 ml) milk. Bake in a slow oven (300°F/150°C/mark 2) for half an hour, or until the custard sets.

PAPAYA FOOL

Take a very ripe papaya and peel it. Cut in half and scoop out the seeds. Slice up the flesh and mash it thoroughly, then rub through a sieve. Add thick coconut cream (see recipe for breadfruit with coconut, p. 59) until the right consistency is obtained, stirring in castor sugar to taste. Flavour with the juice of a lime, and serve in glasses, chilled.

NAMES: Carib. *ababai*; Fr. *papaye*; Sp. *papaya*; Brazil *mamao*; Cuba *fruta bomba*; Mexico *melon zapote* ('tree melon').

STORAGE: At room temperature until ripe, then refrigerated for a short period.

Passionfruit

THIS is the fruit of various passion-flowers *(Passiflora)*, more widely known for the extravagant flower whose parts have been seen as semblances of the Passion of Christ. The climbing plants, also known as Granadillas, are native to South America, but grow in the Mediterranean, in Australasia, and similar climates. The Giant Granadilla, *Passiflora quadrangularis*, is the large vine grown both for ornament and also for dessert fruit; its thick-skinned ovoid fruit, as the Spanish name – *granadilla* – implies, is a 'little pomegranate', having numerous black seeds set in pulp. *P. edulis* is South African and Australian, with the more familiar globose or egg-shaped fruit, thinner-skinned, with juicy, aromatic pulp round each pip, both sweet and acid, and of good flavour. As the fruit ripens, it turns from mauve-red to purple-brown, and the skin to a hard, wrinkled cover. In this form it is usually sold, and although it is useful in many recipes where the pulp can be employed, it is also excellent to eat spooned out of a halved fruit, needing only a touch of whipped cream. The West Indies have their own passionfruit, *pomme d'or*, which is *P. laurifolia*. Venezuela has a sweet granadilla, *P. ligularis*.

GRANADILLA JELLY

This recipe must use large granadillas, not the more readily available small passionfruit. Halve 4 granadillas, and slice up 2 lemons. Place in a pan, cover with water, and cook steadily for 2 hours before straining through a muslin cloth. Then add 2½ lb (1¼ kg) sugar, dissolve over a low heat, and cook again until the mixture turns to jelly, after about 2 hours. Leave it to cool, then pot.

PASSIONFRUIT FLUMMERY

Dissolve ½ oz (15 g) gelatine in hot water, whip ½ pint (300 ml) unsweetened evaporated milk until frothy, add the dissolved gelatine, and heat gently to blend it in. Chill, and the mixture will thicken. Now fold in the pulp of 6 passionfruit, add castor sugar as needed, and pour into a mould. Chill further, until set. Turn out of the mould and serve with cream.

PASSIONFRUIT CREAM

Take 1 small papaya, remove all the seeds, and place the flesh in a bowl. Whip this up until it foams and, separately, whip $\frac{1}{2}$ pint (300 ml) double cream and fold it in. Now fold in the flesh of 4 passionfruit and add $\frac{1}{2}$ teaspoon vanilla essence and 1 teaspoon ground mace. The sweet should be firm; chill before serving.

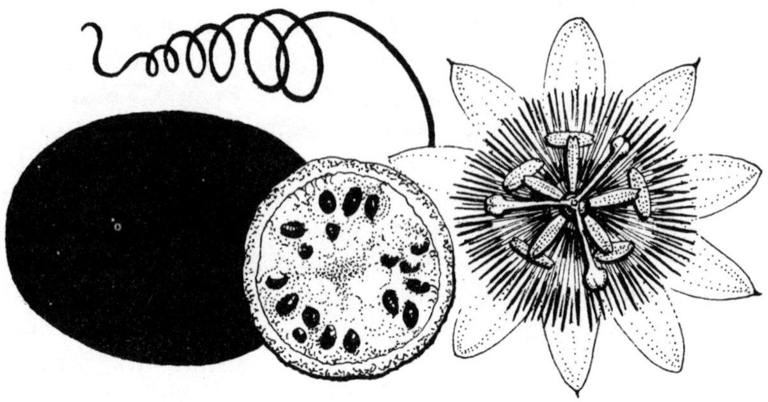

PASSIONFRUIT PIE

Take an 8-inch (20-cm) pie dish and line the sides and bottom with biscuit crumb pastry: to make this mix 3 oz (75 g) melted butter and $1\frac{1}{2}$ tablespoons sugar with $\frac{1}{2}$ lb (225 g) biscuit crumbs – digestive biscuits are ideal. Leave to chill and set while the filling is made. For this, soften 2 teaspoons gelatine in 5 tablespoons water. Add the juice of 1 lemon to 3 egg yolks, beaten to a froth; stir and cook in a double saucepan until the mixture thickens. Add the gelatine, and stir, off the heat, until it is dissolved, then add the pulp of 12 passionfruit. While the mixture is cooling, beat the 3 egg whites until stiff, add 4 oz (100 g) sugar and beat until thick. Now fold the egg whites into the passionfruit mixture and pour into the lined pie dish. Chill in the refrigerator until the pie has set. Serve with whipped cream.

PASSIONFRUIT TEA CAKE

Cream ½ oz (15 g) butter and 2 tablespoons castor sugar, add 1 well-beaten egg and ½ teaspoon grated lemon rind, then beat in the pulp of 2 passionfruits. Fold in 4 oz (100 g) sifted self-raising flour and 5 tablespoons milk and mix to a smooth batter. Turn into a greased 6-inch (15-cm) tin and bake for 20 minutes in a moderate oven (350°F/180°C/mark 4). After 10 minutes' baking, sprinkle the tea cake with brown sugar – 1 dessertspoon will suffice. Return to the oven, finish the cooking, and serve hot with butter.

NAMES: Giant Granadilla: Fr. *barbadine*; Sp. *granadilla*; Brazil *maracuja-assu*.
Granadilla: Fr. *pomme liane*; Venz. *parcha*; Dutch *eiervrucht*.

STORAGE: When ripe passionfruit will keep a day or two in the refrigerator.

Peach

THE emblem of immortality and a token of friendship, the peach, *Prunus persica*, is a Chinese fruit, cultivated in China for many centuries with little change, and pervading many oriental traditions. The fabled golden peaches of Samarkand never left the imperial orchards at Changan, but eventually the caravan routes through Kashmir bore the peach tree to Persia, and from there the Romans took it to Italy. Among the most attractive of blossoming trees, the peach is small and sometimes short-lived, producing fruit which varies in its season, its size, and the colouring of its skin and flesh. The velvet down on the skin is always there, but the skin itself varies from white tinged with green to a golden yellow, with a crimson blush that is anything from a suggestion of colour to a whole covering. The flesh may be white, greenish or yellow, the white being generally considered the better flavoured and more suitable as a dessert fruit, the yellow for canning or cooking. The stone may be free, or may adhere to the flesh, the cling-stone variety.

Many peaches graced Elizabethan orchards and gardens, a novel fruit then, but considered much inferior to the smooth nectarine, and growing only on south-facing walls, in sheltered spots, or under glass. All temperate climates eventually grew peaches, and much skilled attention has been given to the fruit in France, the peaches of Montreuil-sous-Bois being renowned. British peaches were much improved by Thomas Rivers in the nineteenth century, but their cultivation remains largely amateur, and their consumption in Britain relies on the two major exporting countries, Italy and America. It is probable that the peach was introduced to Central America by the colonizing Spaniards, but some impetus came from the North: when the governor of the Massachusetts Bay Colony ordered peach stones from England in 1629, he unleashed a tree that is raised with such facility there that less than 20 years later an observer wrote, 'Peaches better than apricocks by some doe feed Hogs, one man hath ten thousand trees.' Their abundance caused them to run wild and the naturalist Bartram concluded, in the eighteenth century, that the peach was a native American fruit. It nevertheless continued to be cultivated, commercially in California to a huge extent, domestically in many varieties (Hedrick in his *Peaches of New York* records over 2,000 sorts).

The peach has always been ideally a dessert fruit to be eaten out of hand; it occurs somewhat scantily in standard French cuisine in tarts, bavarois and compotes, in *canard aux pêches*, and in a number of chilled desserts, its succulent flesh and inviting aroma demanding no more than that it should be consumed with as little alteration as possible.

LADY SYSONBY'S PEACHES

Cut fresh peaches in quarters and serve up dusted with sugar in fresh champagne, with rose petals floating on top.

PÊCHES AU VIN BLANC

Soak a number of peeled peaches in white wine for half an hour, then drain off and reserve the wine. Fry the peaches very gently in butter, and place them in a covered, warmed dish. Take the butter left in the pan, and thicken it with a little flour to make a sauce, adding the wine the fruit was soaked in. Serve the peaches hot, pouring the sauce over them.

PÊCHES À LA GRECQUE

Using 1 lb (450 g) sugar and 1 pint (600 ml) water prepare a sugar syrup that is not too thick and add a little vanilla; when it is boiling, drop in 6 perfectly ripe peaches, removing them after 2½ minutes. The skins will slip off easily, without disturbing the flesh of the peaches. Return the peaches to the syrup for 5 minutes' simmering, then remove them. Now add more sugar to the syrup, gradually letting it thicken until it is quite heavy. Take the peaches out and arrange round a bowl, leaving a well in the centre. Fill this with apricot jam. Drench the peaches with a sufficiency of the vanilla syrup, to which a liberal amount of curaçao has been added. Prepare enough thick, whipped cream to cover the peaches, adding 1 tablespoon curaçao to this. Chill the bowl of fruit well, and before serving mask the peaches with the cream, and sprinkle a dusting of finely chopped pistachios and hazelnuts on top of this.

PÊCHES FLAMBÉES

Take 6 fresh peaches, place them in a saucepan, and cover them with boiling water. After a few minutes, remove and peel them. Then poach in ¾ pint (450 ml) water for 3 or 4 minutes, over a low flame. Place in a chafing dish, add 2 oz (50 g) sugar and 5–7 fl oz (150–210 ml) peach brandy – or simply brandy – and bring to the table. Light the flame under the chafing dish. As soon as the syrup begins to boil, light it and pour it over the peaches, serving each peach lighted. Cream, lightly beaten, may be served separately.

GINGER PEACHES

Select 6 large, ripe peaches and cut them in half, carefully removing the stones. Arrange them on a baking pan and pour on each half peach a little ginger syrup from a jar of preserved ginger. Bake in a moderate oven (350°F/180°C/mark 4) until the peaches begin to dry and turn slightly brown, then remove and cool. For each serving, place 2 halves side by side, fill their centres with plain ice cream, and pour a little more syrup over them. Finally, sprinkle with finely chopped preserved ginger and chopped walnuts.

PÊCHES CARDINAL

Heat ½ lb (225 g) sugar and 1 teaspoon vanilla essence in just over 1 pint (600 ml) water until the sugar dissolves, and leave the syrup simmering. Place 5 ripe peaches in the syrup, and keep it simmering for 5 minutes; remove from the heat, and let the peaches cool in the syrup. Then drain the fruit, which should still be warm, peel it and arrange on a dish. Set aside to chill. Now make a purée with ½ lb (225 g) ripe raspberries and blend in 4 oz (100 g) sugar, beating until the sugar dissolves. Chill the raspberry purée, and pour over the peaches when serving. Finally, sprinkle with slivered almonds.

TARTE AUX PÊCHES

Prepare a 8–9-inch (20–22½-cm) shortcrust pastry shell (see p. 13), sweetened by adding 1 tablespoon castor sugar to the flour, and partially bake it blind. Take 4 medium-sized ripe peaches and drop them for ¼ minute into boiling water, then peel them, halve them and remove the stones. Sprinkle sugar thickly on the base of the pastry shell and arrange the fruit over it, either in halves, flat side down, or overlapping in a circle if further sliced. Dot with butter and sprinkle a little more sugar over the fruit before baking for 30 minutes in a moderate, preheated, oven (350°F/180°C/mark 4). Finish with an apricot glaze made from sieved apricot jam and a little water and 1 oz (25 g) finely slivered almonds; if desired, caramelize with sugar under a grill and flambé with 2 tablespoons kirsch on serving.

As nearly as can be ascertained, this is Escoffier's original recipe created to honour Dame Nellie Melba at the Savoy in 1892:

PEACH MELBA

Fresh peaches are essential. Peel them, halve them and remove the stones. With ½ lb (225 g) sugar to 1 pint (600 ml) water, make a syrup, and when it is boiling gently add the peaches. Cook until tender, but lift out and chill while the fruit still retains its shape. Reduce the syrup until it thickens to half the original quantity, and to ½ pint (300 ml) syrup add 2 tablespoons prepared raspberry purée; mix and chill. When serving, cover 1 peach half with a generous amount of ice cream, pour over raspberry syrup, and cover with another half peach. Decorate the top with whipped cream and finely chopped nuts.

NAMES: Fr. *pêche*; Ger. *Pfirsich*; It. *pesca*; Sp. *melocoton*, *persico*.
STORAGE: Peaches ripen at room temperature, and should then be eaten; deterioration after ripening will be rapid.

Pear

THE pear *(Pyrus communis)*, a European and Asiatic fruit, has been cultivated since time immemorial, although the Romans were the first to devote themselves to its development. In temperate lands, the pear tree flourished. Under the Medici, over 200 sorts of pears were available at the table of the Grand Duke Cosimo III, and in England, by 1842, the Horticultural Society's gardens had proved more than 700 distinct sorts. About this time, on the banks of the

Loire, a pear germinated and in due course blossomed, yielding in 1849 the first Doyenne de Comice, the first perfect combination of flavour, texture and aroma in the history of the pear. But England had its successes too. There is, in Aldermaston, a churchyard stone to the parish schoolmaster, John Stair, who raised a pear first called Stair's Pear after its appearance in 1770, and now known as Williams after a London nurseryman, and as Bartlett after a similar American entrepreneur. Since the eighteenth century it has become a universal pear, although less grown in Britain than the parthenocarpic Conference pear, winner of the 1885 International Pear Conference.

The cultivation of pears has always – at least since the eighteenth century – tended towards an ideal dessert fruit; however, there are varieties (Yellow Huffcap, Red Pear, Taynton Squash) raised exclusively for perry and, in Normandy, for fermentation into calvados, and a number of trees which produce fruit only good for slow stewing or baking. Jargonelle is an old variety of 'stewing' pear, not often seen now; Pitmaston Duchess, Beurre Clairgeau and Bellissime d'Hiver are other kinds of hard pear. Many early recipes, from mediaeval times onwards, use a generic term, 'wardens', for cooking pears, and invariably refer to them as an alternative to quinces.

Pears are, as a whole, further classified by shape: round or flattened, Bergamot (top-shaped), conical, pyriform and waisted, oval, or calabash (the long, green cooking pears) – but all this is of little avail in the kitchen, where the seasons restrict the choice and it is a matter of experience to distinguish those pears that can be employed for both dessert and cooking, as well as those that should be reserved for the brief period when they will be perfect for eating. A russet fruit, Passe Crasanne, is a good versatile pear, as is the Conference. Doyenne de Comice, although connoisseurs would prefer it to ripen to a buttery, melting flesh and be eaten with nothing more than a little Rocquefort, can be cooked to advantage. Catillac is an old-fashioned sound cooker with flesh that turns red; Docteur Jules Guyot is a sizeable midsummer pear with agreeably juicy flesh, and many other varieties commend themselves to different tastes. In America the miniature Seckel, the graceful Bosc and the Anjou are among the most favoured pears.

No fruit is so thankless as an unripe pear, which one suspects may be a pear that will not ripen. Tusser rightly says that pears gathered too soon will taste of the wood, will shrink and be bitter. But, miraculously, cold and iron-hard pears do after days of waiting turn into fragrant, *fondant* delights. The temptation is to use pears too soon. Deciding when a pear is ripe, or how long it will take to ripen in a fruit-room, or its substitute, or at room temperature, is not easy. As pears ripen from the inside outwards, they should not be kept until the outer surface is soft; they would then be too soft, sleepy and tasteless. Pears, when ready to eat, generally develop an aroma; the flesh near the stem begins to yield, as does that near the base. However, weak flesh near the stem when the pear is picked or purchased indicates immaturity, fruit that will not ripen. Pears stored while they await ripening should not touch. In the cooking of pears, the drawback of a rather uninteresting pallor can be overcome by a judicious use of cochineal or, better still, red wine – fortunately many pear dishes involve stewing or poaching. Pears for cooking should be slightly less than ripe; fully ripe, they are better employed in cold desserts.

PEAR SOUP

Peel and core 2 lb (1 kg) pears that are not yet fully ripe, cut them into pieces and place in a pan. Cover with 2 pints (1·2 litres) water, and cook until tender. Sieve the fruit and set the purée to one side, keeping the juice in the pan. Add 1 tablespoon potato flour, blended with a little water, to thicken it, and bring to the boil. Then add the juice of 1 lemon, stir in a spoonful or two of castor sugar to taste, and add 2 tablespoons white wine before stirring it into the fruit in a dish. When serving, sprinkle flaked almonds over the soup.

BACON, PEARS AND POTATOES

Boil 2 lb (1 kg) bacon for 45 minutes. Melt 2 oz (50 g) butter in a saucepan, add 1 oz (25 g) sugar and stir until browned. Peel and quarter 1 lb (450 g) hard cooking pears, removing the cores, and brown them in the saucepan for a few minutes. Cut 1 lb (450 g) peeled potatoes into pieces about the same size as the pear quarters, boil in salt water for 5 minutes, strain, and add to the pears. Place the bacon on top of them, pour 1 pint (600 ml) stock over it, cover, and cook for 45 minutes, until almost dry. When serving, allow an equal mixture of pears and potatoes; slice the bacon finely and place it on top.

BAKED PEARS

Choose a number of firm 'cooking' pears of equal size, peel them down from the stem end nearly 2 inches (5 cm), slice off the tip and remove the cores with an apple corer. Place each pear in a buttered, ovenproof dish with a lid. Fill the hollow cores with a purée made of stewed apricots or fresh figs, and add 2 teaspoons water and some lemon juice to each ramekin. Butter the tip of the pears, replace the lids on the dishes, and bake the pears until tender, about 1 hour in a moderate oven (350°F/180°C/mark 4). Serve hot with cream.

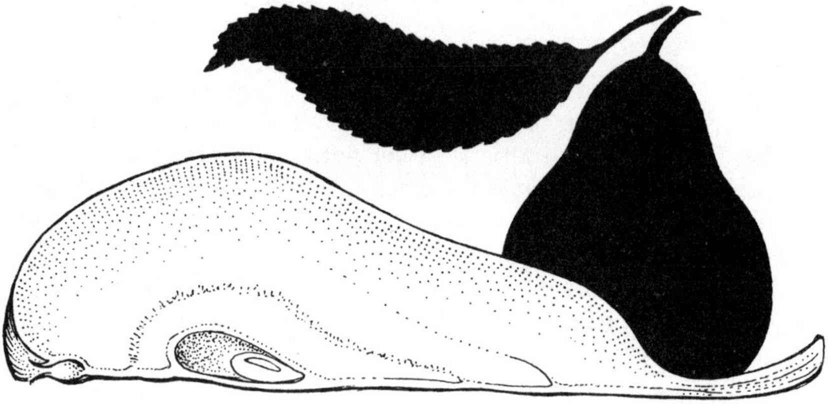

Miroton is strictly the French for a stew of yesterday's beef; here it is applied to pears. The recipe could equally well be adapted to apples:

MIROTON OF PEARS

Peel a number of pears and cut them into quarters, removing the cores. Cook them in water with 1 tablespoon sugar for 30 minutes. Melt some butter in a saucepan, add 1 tablespoon flour and cook together gently. Add the water in which the pears were cooked and 4 tablespoons white wine. Boil to a thick syrup. Heat the pears in the syrup and serve on fried *croûtes* (rounds of bread).

PEARS AU BEURRE

Carefully peel and stalk a number of pears, core them, and cut them in half. Lay them in a pie dish and place a nut of butter in the centre of each half. Sprinkle castor sugar over them and bake in a hot oven (425°F/220°C/mark 7) until done. Alternatively, the pears can be sliced and laid in successive layers of pear and butter and castor sugar; the dish is then browned in a hot oven when full.

In this Normandy dish of baked pears, the *douillons*, or casings, resemble the mediaeval pastry 'coffins' so often used to encase fruit:

DOUILLONS DE POIRES

Make some good shortcrust pastry (see p. 11). Roll it out until it is very thin. Peel a number of pears that are firm-fleshed, but not hard, and leave the stalk intact. Roll them in sugar so that as much adheres as possible; sprinkle more sugar on the pastry, and wrap each pear in it. The stalk should be left sticking out at the top, and the pastry should follow the contours of the pear closely. Bake for 15 minutes in a moderately hot oven (400°F/200°C/mark 6), then reduce the heat to moderate (325°F/165°C/mark 3) and cook until the pears are soft.

This is a sixteenth-century recipe, meant to be eaten when the pears in the pot have grown cold. The addition of whipped cream when serving, and sugar as necessary, will enhance the golden brown pears:

PEARS AND VINE LEAVES IN CIDER

Put a layer of vine leaves in an earthen pot, and lay over them peeled and halved pears, with the cores removed. Add more leaves and pears again until the pot is full; with each layer of pears add some finely sliced ginger and a clove or two. Then fill the pot with as much cider as it will hold, placing a dish or board on the pears so that they do not swim; cover the pot closely and stew it 'all night when the fire be low – or in the oven after the bread be drawn' (i.e. several hours over a low flame).

PEARS IN SYRUP

Take 6 or more firm pears that will not ripen, pare them, and halve them and remove the cores if they are large. Place them in a pan with ¼ pint (150 ml) vinegar, 1 pint (600 ml) water, 1 lb (450 g) sugar, 2 cloves, and a piece of bruised ginger root. Heat slowly to dissolve the sugar, then bring to boiling point, cover, and leave simmering until the pears turn dark brown and the juice is reduced to a rich syrup. Check that the pears are done by piercing with a fork. Strain off the juice to serve over the fruit, which should be garnished with whipped cream. A raspberry sauce may also be served with the pears. Serve cold.

PEAR AND GINGER PUDDING

Butter a pudding basin and coat the inside with breadcrumbs. Prepare a bread mixture of 4 oz (100 g) breadcrumbs and 2 oz (50 g) grated suet. Wash, peel and core 1 lb (450 g) pears, and grate them with the rind of a lemon. Add a little lemon juice to prevent discoloration. Add some sugar and 1 oz (25 g) preserved ginger, finely chopped. Place one-third of the bread mixture in the basin and next half the pears, then a third of the bread again, and the second half of the pears; cover all this with the remainder of the bread mixture. Cover with greased greaseproof paper or foil. Steam for 2 hours. Serve hot with custard or an apricot sauce.

STEWED PEARS

Peel 6 ripe pears that are still firm, and cut them in half, removing the cores. Sprinkle with a little lemon juice. In a saucepan, place 2 tablespoons lemon juice, 6 oz (175 g) sugar, ½ pint (300 ml) water, ½ pint (300 ml) red wine and a stick of cinnamon. Stir to dissolve the sugar, bring to the boil, then add the pear halves and let them simmer for at least 15 minutes, until the fruit is tender. Place the fruit in a serving dish, leaving the liquid in the pan. Add 2 slices of lemon to this, cook until it is syrupy, then strain and leave to cool. Pour over the pears. Serve the whole dish cold, with whipped cream.

POIRES BELLE HÉLÈNE

Peel and halve 4 pears, remove the cores, and poach gently for 15 minutes in a syrup made from $\frac{1}{2}$ lb (225 g) castor sugar and $\frac{1}{2}$ pint (300 ml) water. Break up 6 oz (175 g) plain, bitter chocolate into a saucepan; add 1 oz (25 g) butter and 2 tablespoons water. Melt very gradually, stirring with a wooden spoon until a smooth creamy sauce forms. Add a little vanilla essence and 4 tablespoons double cream; stir in thoroughly. Serve 2 halves of pear in a glass with a ball of vanilla ice cream as their core; pour hot chocolate sauce over the pear, and serve at once.

NAMES: Fr. *poire*; It. and Sp. *pera*; Ger. *Birne*.
STORAGE: See above.

Persimmon

PERSIMMON is an Algonquin Indian word for the fruit of *Diospyros virginiana*; it grows wild in America, and was first reported to the Old World by Captain John Smith as 'putchamina'. There was some confusion with the medlar in its naming by early observers, little clarified by its being also known as the Virginia Plum and then the Virginia Date. The fruit is indeed the size of a large plum, varying in colour from green to yellow, and from orange to purplish-red, even a dark red in northern States when the fruit shrivels on the trees and approximates to the date in appearance and, to some palates, in taste. A bread and a beer are made from the fruit; it needs to be fully ripe before it is sweet enough to eat out of hand, the tannic acid in green fruit giving it an unacceptable astringency.

Some further ambiguity arises from the general use of the name persimmon for the fruit of another and distinct species of Diospyros, the Japanese *D. kaki*. This itself originated in China, but

the Japanese have devoted much attention to producing hundreds of varieties, and consider it the best of their fruits. It has been cultivated on the Côte d'Azur, and in Italy since 1884, and was introduced into the United States by Commodore Perry's expedition when the U.S.S. *Lexington* returned in 1855 with trees that were planted at the Botanic Garden at Washington. Regular culture of the Japanese persimmon had started there by 1870. In England, it has been grown as a wall tree – *D. kaki* is a hardy tree – but the fruit has to be gathered to ripen indoors, protected from frost.

The fruit itself is unmistakable: it is a bright glossy orange, of brittle and slightly waxy appearance, easily bruised or discoloured, and almost without odour; a four-part leaf is flattened over the stem end. The flesh varies in astringency as the fruit begins to soften, and there are variations in the texture and consistency of different fruits, and indeed within one fruit itself. The seeds in commercial varieties are negligible. Some varieties are crisp to eat, others custard-like or pulpy, even sweet and mawkish. The kaki is mainly eaten as a dessert fruit; it is spectacular as one of the most brilliant of fruits and unusual in making its appearance in autumn and winter. Its food value is good. There are high percentages of protein and glucose, and a large count of vitamin A, about the same amount as in the apricot. The 'kaki' persimmon, which does not feature in oriental cooked dishes, can be stewed, but is not adaptable to cooking except where its pulp can be used in mousses, fools, creams and similar mixtures or where its tomato-like character could be utilized. A persimmon jam would be possible, and an equivalent to apple sauce could be made of the flesh.

PERSIMMON PUDDING WITH BRANDY SAUCE

Take 3 ripe persimmons, wash, stem and quarter them and scoop out the pulp. Blend this, or rub it through a sieve. Cream together 4 oz (100 g) butter and 6 oz (175 g) sugar, and beat in 2 eggs; add 1 teaspoon lemon juice, 2 teaspoons vanilla essence and a generous tablespoon brandy. Add the persimmon, and sift 4 oz (100 g) flour with 1 teaspoon cinnamon, and stir this into the creamed mixture. Purée the mixture in a blender. Add 6 oz (175 g) raisins and 2–3 oz (50–75 g) chopped, mixed nuts. Ladle all the mixture into a greased pudding mould, cover it with greased greaseproof paper or foil, and steam for 2 hours, adding boiling water as needed. Cool for at least 10 minutes before turning out of the mould. Serve warm, with a brandy sauce made by beating together 5 oz (150 g) butter, ½ lb (225 g) castor sugar and 3 tablespoons brandy. This should be made beforehand and chilled until the pudding is served.

KAKIS À L'IMPÉRATRICE

Take 6 firm persimmons, remove their top leaves and carefully cut a hole round each stalk. With the skin intact, remove the pulp, dice it and stir it thoroughly into rice à l'Impératrice. To prepare this, cook 4 oz (100 g) pudding rice in 1½ pints (900 ml) water, flavoured with vanilla, and add 2 oz (50 g) butter, 2 oz (50 g) sugar and, when the rice is cold, 2 tablespoons finely chopped crystallized fruit that has been soaked in kirsch; thicken with ½ pint (300 ml) custard, and finally add ½ pint (300 ml) whipped cream. Sprinkle the hollowed out fruit with a few drops of kirsch, then fill with the prepared mixture. Serve chilled.

NAMES: *D. Virginiana* U.S.A. Virginia Date, Virginia Plum. Pessimon, Persimmon, Piakmine; Fr. *plaquemine*; Louisiana Fr. *ougoufle*.
D. kaki U.S.A. Japanese persimmon; Kaki plum, Kaki fig, Keg fig, Chinese date plum; Fr., It. *kaki*; Ger. *Dattelpflaume*.

STORAGE: At room temperature.

Pineapple

THE regal and architectural pineapple (*Ananas comosus*) is no simple fruit, but a sorosis, a collective fruit consisting of a number of fleshy bracts consolidating into a whole – a development of a conical spike of flowers with a tuft of small leaves on top. It belongs to Central tropical America, and the first European eyes to see a pineapple were those of Columbus's men in Guadeloupe, in 1493. It remained an awesome curiosity for many years to come, typified in the many versions of the painting by Danckerts: 'Mr Rose, the Royal Gardener, presenting to King Charles 2nd the first pineapple raised in England.' This was in 1661, and the fruit was a Barbados queen pine. At about this time, John Worlidge wrote: 'In Jamaica and Brasilia, grows the fruit *ananas* on a stalk of a foot long, surrounded with sixteen sharp leaves, between which is the fruit like a pineapple [i.e. shaped like a pinecone] but much bigger; the innermost pulp whereof melts on the tongue, and is of so delicious a taste, that it exceeds all other dainties.' But it was the eighteenth century before pineapples were grown seriously on English hot-beds and in the greenhouses of the great estates. In 1777, Captain Cook introduced the pineapple to the Pacific Islands, and cultivation, from the nineteenth century onwards, was intensified in favourable areas. Hawaii is exhaustively farmed; South Africa, Cuba, Florida and California, Central America, the Azores and the Canary Islands are also important providers of the pineapple in its many varieties, from huge orange fruits to small green Ivory Coast pines, with flesh ranging from deep yellow to white.

Smaller fruits are generally better in flavour, but shorn of the leaves, the outer armour, and the core, the edible part of a pineapple represents only about half of the original bulk. This may seem an argument for tinned or preserved pineapple, but the process destroys the enzyme that makes pineapple a tenderizer of meat and an agreeable companion to grilled beef steaks. In buying pineapples, ripeness is properly determined by sounding, a dull solid sound, and not a hollow one, being desirable, with a very slight yielding to the touch.

This is a Russian salad, *salat iz ananasa s sel'dyeryeyen*:

PINEAPPLE AND CELERY SALAD

Chop up fresh celery and fresh pineapple in equal quantities, and arrange on a bed of lettuce leaves. Use a vinegar and oil dressing, and, as a garnish, slices of cooked beetroot.

PINEAPPLE AND FISH CASSEROLE

Cut the flesh of a small fresh pineapple into cubes, drain them of juice, and place in the bottom of a well-buttered casserole. Spread over the fruit 1 lb (450 g) cooked, flaked fish. Have ready cooked and mashed 1 lb (450 g) potatoes, and beat into these 1 egg, a large pat of butter, 2 tablespoons milk, and cayenne pepper to taste; when fully creamed, cover the fish completely. Bake for 30 minutes in a moderate oven (350°F/180°C/mark 4), remove, grate over the dish a covering of strong Canadian cheddar cheese, and return to the oven. Melt and brown the cheese before serving.

MALAYSIAN CHICKEN WITH PINEAPPLE

Have ready 1 whole boiled chicken, cold; bone it, and cut the white meat up into pieces about 1 inch ($2\frac{1}{2}$ cm) long. Reserve the stock in which the chicken was cooked. Into 2 teaspoons soy sauce stir 1 teaspoon cornflour, add salt and pepper and a little oil, and mix all this with the chicken. Heat some frying oil – groundnut or olive oil – and fry in it half a clove of garlic and the pieces of diced chicken. Have ready a small pineapple, cut into inch ($2\frac{1}{2}$-cm) cubes, and add half of these to the chicken, covering the pan and letting the meat simmer for 5 minutes. Pour off some pineapple juice (from the pineapple cubes) and mix in another teaspoon of cornflour to make a paste. To $\frac{1}{4}$ pint (150 ml) of the chicken stock, add a few slices of root ginger, the rest of the pineapple juice, and the cornflour paste, and cook slowly until a thick sauce results. Serve the chicken and pineapple mixture on an oval dish, with the hot sauce poured over it and the remaining pineapple cubes on top. Serve rice separately.

HAWAIIAN LEG OF LAMB

Have a leg of lamb boned, and then weigh it to determine roasting time, allowing 30 minutes per pound (450 g). Rub the cut rind of a lemon over the lamb some time before roasting, and brush it over with the juice of a lemon. To prepare the stuffing, fry lightly, for a few minutes, in melted butter a crushed half clove of garlic mixed with 1 tablespoon scraped onion pulp, then add 2 tablespoons finely chopped parsley, ½ teaspoon ground ginger, salt and pepper to taste, 1 tablespoon brown sugar and, lastly, the diced flesh of a small fresh pineapple. Mix with 6 oz (175 g) toasted breadcrumbs, and fill the bone cavity with the stuffing, sewing up afterwards. Brush the leg with 2 tablespoons melted butter, and dust lightly with salt, pepper and a little ground ginger. Brown the leg in a moderately hot oven (400°F/200°C/mark 6) for a short while, then reduce the heat to moderate (350°F/180°C/mark 4), roast, and baste with pineapple juice every 10 minutes. Thicken the pan gravy when the leg is done, and serve as a separate sauce.

PINEAPPLE FRITTERS

Pare and core a ripe pineapple. Cut into slices and stew with a little water, sugar and lemon peel. When soft, add a little white wine and the juice of half a lime. When cold, make a batter of 4 oz

(100 g) flour, 2 tablespoons cream, 4 tablespoons white wine, some sugar and 4 eggs. Beat it all very well together. Put a little butter into a frying pan, coat the fruit with the batter, and spoon the pieces out one by one to fry to a light brown. Place on a sieve to dry, and serve hot, with plenty of castor sugar.

PINEAPPLE FLAN

Bake a flan case (see p. 13) and let it cool. Arrange on it thinly sliced rings of fresh pineapple so that they overlap in the same direction in a circle. Make a sauce by boiling some pineapple juice (from cutting up the fruit) with 3 tablespoons apricot purée, the juice of 1 lemon, 1 teaspoon cornflour and 1 tablespoon rum. After boiling, strain the sauce, cool it, and pour it over the slices. Cover the whole flan with whipped cream, and sprinkle on top some finely chopped pistachio nuts.

PINEAPPLE SOUFFLÉ

Cut up into fine pieces the segments of a small tin of pineapple slices. Mix this with the yolks of 3 eggs, adding a little sugar. Put this mixture into a buttered soufflé dish. Beat the whites of 6 eggs to a stiff froth and fold them in. Cook in a moderate oven (350°F/180°C/mark 4) for approximately 20 minutes.

ANANAS AU KIRSCH

Cut a whole pineapple in half lengthwise, and scoop out the fruit. Keep the shell. Cut the fruit into small cubes, no more than ½ inch (1 cm) across. Add castor sugar to the fruit as necessary, and a liberal helping of kirsch, and replace the mixture in the shell. Cover with whipped cream. Decorate with 4 oz (100 g) mixed fresh raspberries and wild strawberries.

PINEAPPLE CREAM

Cut a fresh pineapple into slices, and make a sandwich of every pair with a filling of whipped double cream beaten with castor sugar and flavoured with vanilla. Sprinkle with a few drops of rum, then add more cream on top and garnish liberally with finely chopped walnuts.

PINEAPPLE DELIGHT

Sever the leafy top from a small, ripe, well-shaped pineapple. Cut round inside to loosen and remove as much of the flesh as possible without piercing the skin. Remove the core from the pineapple flesh, cut into chunks, place in a basin, sprinkle lightly with rum and leave to chill. Treat the inside of the pineapple to a sprinkling of rum as well, and add a crushed sprig of mint. Replace the top and chill. When ready to serve, remove the mint and fill with the pineapple chunks, to which ice cream has been added. Serve with the top replaced.

PINEAPPLE À LA CREOLE

Cut in half a medium-sized pineapple, and scoop out all the fruit. Mash with 2 forks until it forms a pulp. Take 3 bananas and 3 oranges, peel them, slice the bananas into rings and mash the flesh of the oranges, first removing each segment from the surrounding membrane. Add to the pineapple pulp. Grate a small coconut finely and mix it into the other fruit; sharpen with the juice of a lemon and sweeten with sugar to taste. Chill in a refrigerator for a number of hours before serving very cold.

NAMES: Fr. *ananas*; Sp. *pina*; Brazil. *abacaxi*.
STORAGE: A ripe pineapple should keep a week in the refrigerator.

Plum

THERE are numerous wild plums, cooking and dessert plums, all stretching geographically between the Pacific coast of America through Europe to Japan, ranging from the bitterest sloes to the choicest golden wall-trained fruits. Under the general classification of *Prunus domestica* are plums of great diversity in shape, size, flavour and colour; black, purple, deep red, golden, green and yellow skins cover flesh that, in cooking plums, tends to be juiceless and acid, and in dessert varieties, sweet and melting. The most famous of these, the red Victoria, was found in a Sussex wood about 1840, and has proved a hardy, heavy-cropping variety and an important orchard fruit.

But it is the closely related members of the plum family, the gages, damsons and sloes, that are most attractive in the kitchen, for even the best cooked plums seem to lack verve, and are perhaps overshadowed by memories of stewed plums, of that dour jam, plum and apple, and of dreary fruit eaten before it is ripe. But, as Culpepper said, plums are under Venus, and are like women – some better and some worse. Damsons are excellent for pies and open tarts – a damson flan over which a lattice of cream cheese has been piped is a compelling dish – in spite of the stones (which should not be removed before cooking), greengages – the French *reine-claude* – are good for dessert or cooking, and, for eating out of hand only, the mirabelle is highly prized. The bullace and the sloe find their place in cheeses, pickles and preserves, the sloe in particular being commendable in sloe wine and sloe gin. Indeed the whole genus donates handsomely to the province of drink: there is *quetsch*, the plum brandy, *prunelle*, the French sloe-based liqueur, *mirabelle*, also a French spirit, the Japanese plum wine *umeshu*, a sort of ratafia of green plums matured in sugar and spirit, and the excellent Serbo-Croatian *slivowitz*.

One constant disappointment in any search for plum recipes are the misleading plum puddings, plum duffs and plum porridges that persist in old cookery books. At a very early stage in British cookery the plum in these dishes, probably small and hard, was replaced by the sultana and the raisin. A seventeenth-century recipe for plum porridge includes currants, raisins and prunes – at least a form of plum. The following recipe, prepared for King George I to eat on Christmas Day 1714, must be one of the last plum puddings still to use plums:

CHRISTMAS PLUM PUDDING

Mix 1½ lb (675 g) shredded suet, 1 lb (450 g) each of stoned and halved plums, shredded mixed peel, raisins, sultanas, currants, sugar, sifted flour and breadcrumbs, 1 heaped teaspoon mixed spice, half a grated nutmeg, and 2 teaspoons salt. Add 8 eggs that have been beaten to a froth, ½ pint (300 ml) fresh milk, the juice of 1 lemon, and 5 tablespoons brandy. Stir well. Leave in a cool place for 12 hours, then turn the mixture into 3 buttered moulds and cover with greased greaseproof paper or foil. Boil for 8 hours – the first boiling – and then boil fror 2 hours before serving.

GREENGAGE CROÛTES

Cut a number of greengages in half, and remove the stones. Arrange rounds of bread in a fireproof dish and sprinkle a little castor sugar over them; place halves of fruit, rounded side uppermost, on each round. Dot with butter. Sprinkle more sugar on top. Bake in a moderate oven (350°F/180°C/mark 4) for up to half an hour. Serve hot, with more sugar if necessary, and accompanied by a purée of apricots.

GREENGAGE FRITTERS À L'AGENAISE

Select some fine, ripe greengages and slit them into two to remove the stones. Coat them in castor sugar, steep them in brandy for at least an hour, drain them, and fill each with a spoonful of thick plum purée flavoured with kirsch. Dip them in prepared batter (see recipe for Bundaberg banana fritters, p. 44), and fry in smoking hot deep fat. Drain the fritters when ready, sprinkle with castor sugar, and serve piled on a napkin-covered dish.

DAMSON CHEESE

Set the damsons in a pan with a little water. Simmer gently until the fruit is soft and pulpy and the skins are thoroughly softened. Then stir the fruit and press through a sieve. Add 1 lb (450 g) sugar for every pound of pulp, put the damsons back on the fire in a pan, dissolve the sugar over a low heat, stir well and simmer for about 2 hours. Then boil quickly for an hour, stirring constantly until a really thick consistency is achieved. Put the cheese in warm jars, place brandied paper over the top, and cover tightly when cold. Keep in a cool place.

A country recipe of long standing, the cheese should be a purple-black in colour. Once made, it should not be touched for six months, and is said to improve for up to two years. When the cheese begins to shrink from the sides of its jar, it should be turned out; it was the practice to eat it as a dessert, stuck with split almonds and doused with port. Much damson and bullace cheese was formerly made in Sussex.

PLUM CHARLOTTE

Stone 12 plums, and just cover them with water in a saucepan, adding enough sugar to sweeten. Stew the fruit until tender. Line a mould with thin strips of sponge cake, and fill with the plums while they are still hot. Cover in with more sponge cake, and leave to cool. Then refrigerate, serving the charlotte ice-cold, with thick cream.

PLUM DUMPLINGS

Dissolve ½ oz (15 g) yeast in ½ pint (300 ml) lukewarm milk. Let this stand for 30 minutes. Stir 1 oz (25 g) melted butter into the yolks of 3 eggs, then add 1 lb (450 g) flour, a little salt and the dissolved yeast. Work continuously, adding the whites of the 3 eggs and extra flour if necessary to make a dough, then let the mixture rise. Remove the stones from the plums and replace with a lump of sugar. Next make little balls out of the mixture, wrapping a fresh, ripe plum into each and coating it well. Stand them to rise. Bring water to the boil, and plunge the dumplings in for 5 minutes each.

When ready, place on a dish, pour melted butter over them, and serve sprinkled generously with castor sugar.

TURKISH PLUM DESSERT

Take a number of large black plums, clean them, and remove the stones via the stem end. Grind 2 oz (50 g) sweet almonds and fill the plums with them, first moistening the almonds with a little rose water. Then take ¼ pint (150 ml) Marsala, 5 oz (150 g) flour, the grated rind of half a lemon and 2 teaspoons castor sugar and whisk into a batter. Let this stand for 1 hour, then whisk again, folding in 2 egg whites which have been beaten until stiff. Melt in a frying pan 6 oz (175 g) butter. Coat the plums with the batter and fry them; they will puff up to a golden brown and can then be drained and rolled in a mixture of icing sugar and very finely chopped pistachio nuts. Serve with whipped cream while still hot.

NAMES: Fr. *prune*; Ger. *Pflaume*; It. *prugna*; Sp. *ciruela*.

STORAGE: Once they are ripe, plums will generally need to be used within two or three days: bruising or damage soon causes deterioration.

Pomegranate

THE pomegranate – the Roman *pomum granatum*, an 'apple with seeds', and Linnaeus's *Punica granatum* (the only genus of its family, the Punicaceae) – has its many seeds in four two-storied compartments, each seed being covered by a sac of red juicy pulp (the aril). The fact that it is highly thirst-quenching, combined with an exceptional keeping quality – the pomegranate's skin simply becomes dull, tougher, and more leathery – has assured the fruit an eminent place in the arid lands of the Near East. The best fruit, with the sweetest juice and the smallest seeds, is said to have been found in

Kabul, and in Arabia Sir Richard Burton found delicious varieties. In fact, the pomegranate is cultivated from Asia Minor to northern China; it also grows in America.

From antiquity the pomegranate has been revered in Persia, Egypt, Greece and Rome, its attractive shape, crowned with

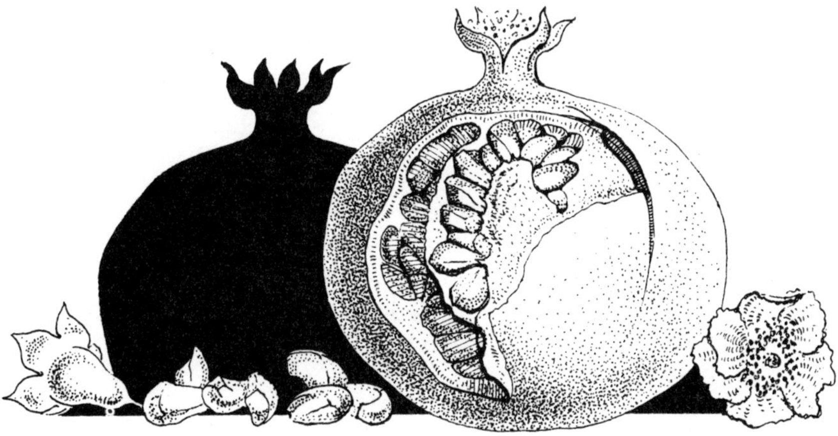

vestigial calyx lobes, and fine colour – deep red, yellowish-pink, with violet-pink sacs – being no small part of its appeal. A mystic fruit, emblem of increase and abundance, it was a prime symbol of Ibritz, the Hittite God of agriculture; it was the badge of Katherine of Aragon; it remains part of the blazon of the Royal College of Physicians, no doubt because of its association with fertility. The fact that the fruit scatters its seeds caused a new metal-scattering hand-bomb to be called a grenade in about 1590, and the Spanish town of Grenada took its name from a similarity in design to a half-opened pomegranate. The excellent French syrup *grenadine* derives from the fruit. There is, however, little to be achieved with the pomegranate in the kitchen. While not being so ungenerous, as some have been, as to suggest that its only value is to display in a basket of dessert fruits, a uniquely flavoured juice, syrup and jelly are the only products of the pomegranate.

Gelee anar is, in Iran, a straightforward dessert; all you need to know is how to obtain the juice from the fruit. The best way to do this is to put pressure on the fruit, working it over so that the inside is crushed and the sacs release their juice. It will be clear when it is time to make a cut in the skin with a sharp knife to let the juice, *abi anar*, flow. It should always be strained.

POMEGRANATE JELLY

Squeeze a quantity of pomegranate juice, and strain it. Add ½ oz (15 g) dissolved gelatine for each 1 pint (600 ml) juice, set, and serve with cream and sugar.

COMPOTE OF POMEGRANATES

Pick out the largest pomegranates from a selection of fruit, and halve them. Remove all the seeds. Place the halves in a dish, and sprinkle them with rose water. Take some smaller pomegranates and crush them for their juice, and add to this sufficient castor sugar and water to make a syrup. Dissolve the sugar over a low heat, then boil until the syrup thickens; let it cool. When cold, add the seeds from the larger fruit and pour the syrup into the pomegranate halves.

NAMES: Eng. Garnet-apple (the precious stone is named after *granatum*); Fr. *grenade*; Ger. *Granatapfel*; Dutch *granaatappel*; Sp. *granada*.

STORAGE: The fruit will keep for many weeks. Soft patches appear on the skin when it has passed its prime.

Prune

THIS semi-fruit, a fruit at one remove, is at once a neglected, if not despised, part of the British kitchen repertoire and a luxury rarely sampled: shrivelled and stewed prunes on the one hand, and on the other the plump *pruneaux d'Agen* nestling in Armagnac. The prune must have an antiquity parallel with that of the plum, but there is no account of it in Roman cooking and no mediaeval recipes are to be found. But in the opening years of the seventeenth century, we read of whole hogsheads of imported 'Damaske Prunes' (i.e. from damsons) being sold at the grocers and of 'black Bulleis' (bullace) prunes, the first to loose and the second to bind the body. It is this first virtue that has brought the prune into such low repute in English cooking: it may be noted that Bishop Berkeley, in his eighteenth-century eulogy of tar-water, was probably the first writer to commend stewed prunes as part of an invalid's diet. The prune is in fact rich in iron and vitamins A and B, but nursery associations with rice pudding have condemned it in many a mind.

Other countries enjoy the prune differently. In France, the Lot and Garonne valleys produce the best plums and the towns of Agen, Marmande, Tonneins and Clairac the best prunes. Only some varieties of plum will dry without fermenting, and these are traditionally left in the sun on straw for forty-eight hours and then further reduced by five or six hours in a mild oven. Provence produces the *pistole* and the lesser prunes called *brignoles*; Metz and Nancy a prune called *quetsch*; Tours and Bordeaux give their names to local prune products, and the Basses-Alpes yield the *pruneau fleuri*, delicate and uncommon. The dark, rich prunes of Agen are the most esteemed. But the comparatively small output of France has to contend with the products of California; though its prunes are not remarkable, California supplies three-quarters of the world's market. It is ironic that they were introduced to America by a French nurseryman, Louis Pellier, in 1856, after he had settled in the Santa Clara valley. Some are now sun-dried, others are commercially dehydrated; they can be bought in Agen as elsewhere.

The prune, after soaking, is adaptable to game, like pheasant or partridge, and to many meats; a French dish of long standing is *noisettes de porc aux pruneaux* (although Oliver Goldsmith speaks slightingly of what must be the same dish in *She Stoops to Conquer*: 'To men that are hungry, pig, with pruin sauce is very good eating'). It is much used in compotes, can readily be stuffed, and allies itself easily with other fruits, with wine and liqueurs, and even, as few fruits do, with chocolate, as in a Hungarian sweetmeat which consists of a stoned prune wholly coated in bitter chocolate.

England has at least this excellent savoury dish:

DEVILS ON HORSEBACK

Stuff some lightly soaked prunes with chutney, small pickled onions, stoned olives or salted almonds (see p. 15), and wrap each one in a trimmed rasher of bacon. Skewer and grill until the bacon is cooked, and before removing from the heat sprinkle some grated cheese over them. Serve hot on hot buttered toast.

The following recipe suggests an alternative breakfast:

CAISSES À LA FLORENCE

Soak a number of the largest prunes available, 1 or 2 for each person. Remove the stones, and stuff the prunes with smoked, cooked haddock, bones and skin removed, and pounded to a smooth paste with cream and a pinch of cayenne pepper. Fry long pieces of bread in butter, place a stuffed prune on each and heat in the oven.

PRUNE SOUP

Cook 2 oz (50 g) sago in 2 pints (1·2 litres) water until it turns transparent. Cook, in another saucepan, 3 oz (75 g) stoned prunes, $1\frac{1}{2}$ oz (40 g) raisins, and 2 oz (50 g) sugar. When the fruit is cooked, add it to the sago with $\frac{1}{4}$ pint (150 ml) blackcurrant juice. Serve hot, with almond paste balls (see p. 16).

Instead of the usual *sauce à la diable* for this chicken dish, one of Lady Sysonby's inventive recipes has a prune and onion sauce:

POULET À LA CRAPAUDINE

Split a spring chicken in half, open it and flatten slightly. Spread with melted butter; season with salt and pepper and grill both sides until golden brown. For the sauce, take 12 very small onions and put them in a saucepan with a piece of butter, pepper and salt. Cover, and let them cook, agitating occasionally to prevent burning. Then stone 12 prunes that have already been cooked and place them in a sauce boat. Put 2 tablespoons castor sugar into a pan and dissolve over the heat until it forms a light caramel. Add 2 tablespoons best vinegar, and reduce until the caramel is quite dissolved; then divide it between the onions and the prunes, and serve them up in separate sauce boats.

PRUNE FLAN

Bake a flan case of shortcrust pastry (see p. 13). Take 1 bottle large preserved French prunes *(pruneaux d'Agen* or *pruneaux de Tours)* and place in a casserole with 2 tablespoons whisky, $\frac{1}{4}$ pint (150 ml) water and 1 tablespoon brown sugar. Cook for 4 hours in a cool oven (225°F/110°C/mark $\frac{1}{4}$). Then stone the prunes and place them in the flan. Thicken the remaining juice with cornflour mixed with a little water, and when cool pour it over the prunes; decorate with stiff whipped cream.

PRUNE PUDDING WITH SHERRY SAUCE

Wash 1 lb (450 g) prunes and steam for 20 minutes over boiling water, using a colander or steamer. Then remove the stones, and cut up the prunes into fine pieces. Mix 9 oz (250 g) sifted flour with 1 teaspoon baking powder, a pinch of salt, and $\frac{1}{2}$ teaspoon each of cloves, nutmeg and cinnamon. Gently warm 4 oz (100 g) butter, add 4 oz (100 g) sugar, and beat in $\frac{1}{4}$ pint (150 ml) golden syrup and 2 lightly whisked eggs. Combine the dry ingredients by alternately mixing well with additions from 5 tablespoons milk and 7 tablespoons sweet sherry, add the warm butter, sugar and syrup mixture, and finally add the prunes and 2–3 oz (50–75 g) chopped walnuts. Turn into a well-greased mould, cover tightly with greased greaseproof paper or foil, and steam, with water halfway up the mould, for 3 hours, adding water as necessary. Meanwhile, prepare a sherry sauce: beat 4 oz (100 g) butter until creamy, gradually add a pinch of cinnamon mixed with 12 oz (350 g) sugar alternately with 4 tablespoons sherry, and beat until the mixture is stiff and creamy. Serve it as a sauce over the hot pudding.

Here is an Austrian sweet of prunes in batter:

SCHLOSSERBUBEN

Soak any number of prunes in cold water until they begin to swell, remove them, dry, and set to one side. Have ready a batter prepared from 4 oz (100 g) flour, 1 tablespoon sugar, 1 tablespoon olive oil, $\frac{1}{4}$ pint (150 ml) beer and the yolk of 1 egg; finally add the white of the egg, beaten until stiff. Remove the stones from the prunes and stuff with coarsely chopped almonds. Dip the prunes in the batter and fry in very hot fat, drain on paper, and serve sprinkled with grated chocolate.

Also from Austria is this *Nusstorte*, served warm with whipped cream and additional walnut halves on top:

WALNUT CAKE WITH APRICOTS AND PRUNES

Steam 5 oz (150 g) prunes and an equal amount of dried apricots for about 20 minutes. Cut into small pieces. Separate 3 eggs; in one bowl beat the egg whites until they stand in peaks, and in another bowl beat into the egg yolks 4 oz (100 g) granulated sugar, 1 tablespoon milk and 2 teaspoons rum. Fold the whites into the yolks. Sift together 4 oz (100 g) flour, $\frac{1}{2}$ teaspoon baking powder and a pinch of salt and gradually add this to the mixture to make a smooth batter. Fold in the steamed prunes and apricots and 5 oz (150 g) chopped walnuts. Pour the batter into a deep cake tin, lined with greaseproof paper, and bake for about half an hour in a moderate oven (350°F/180°C/mark 4). Allow to cool for 10 minutes before turning out.

A French dessert such as *pruneaux de Touraine* would simply be Tours prunes stewed with a local red wine, sugar and cinnamon. Here is another similar recipe; these prunes in spiced wine could be eaten by themselves, hot or cold, or used as a garnish for cold ham or turkey.

PRUNES IN SPICED WINE

Wash 1 lb (450 g) prunes and soak in warm water until they soften. Simmer for 10 minutes in a saucepan with 1 pint (600 ml) claret and ½ pint (300 ml) water. Add 4 oz (100 g) brown sugar, ½ teaspoon ground allspice and ¼ teaspoon ground ginger. Cook for about 15 minutes until the prunes are plump. Serve in the liquid.

NAMES: Med. Latin *pruna*; Gk. *prunon*; early Eng. proyne, pruine, pruan, etc.; Fr. *pruneau*; Ger. *Dörrpflaume*; It. *prugna secca*; Sp. *ciruela pasa*.

STORAGE: Indefinitely in brandy or Armagnac; some months in air-tight paper bags, polythene bags or stoppered glass jars.

Quince

THE true quince (*Cydonia oblonga*) and its variants seem to have come from the Middle East. Well before the Christian era, the fruit was cultivated in north Persia and Armenia, spreading to Asia Minor; by the seventh century BC the quince, named after Kydonia (Khania) in Crete, was common in Greece. Colonists brought it to Italy, and in time the quince was established throughout Europe. The Anglo-Saxons had quince trees, semi-wild perhaps, but the quince was orchard-grown by the later mediaeval period. But as the old quince orchards of Europe have long since vanished, so in southern England there is now rarely more than a single tree against a sun-catching wall or near a pond.

A shallow-rooted deciduous tree, slow of growth, spreading tortuous branches over damp soils unloved by other fruit trees, the quince is of easy culture and free from pests, living to a great age. In September and October the hard fruit mellows, and should be gathered when dry to ripen. When ripe it is golden yellow, lightly fluffed with a cottony frieze, of irregular shape and heady fragrance. It is not surprising that the 'quick peculiar quince' was, for the classical writers, an emblem of love and happiness, dedicated to Venus and adorning her temples; it was a token at marriage rites in ancient Greece, and in England as recently as the eighteenth century. The ceremonial eating of a quince was to be 'a preparative of sweete and delightfull dayes', and the old herbalists were consistent in recommending the fruit for languid appetites and troubled stomachs.

It is an unusual fruit in that it cannot be eaten raw, but once stewed, baked, scalded or roasted, the austere, almost wooden, flesh miraculously yields a reddish-pink food that few can resist. Earlier ages employed the quince extensively for many marmalades – preceding the orange – for astringent jellies, compotes, pulps, creams, pastes, and the like. Parkinson wrote that there was no fruit growing in the land with so many excellent uses. Yet where John Nott, the Duke of Bolton's cook, has twenty-three recipes for quince – twice as many as for the apple – in his *Cook's and Confectioner's Dictionary* of 1723, the fruit vanishes from Victorian cookery books, emerging in the 1930s with a revival of interest in neglected country dishes. This was not matched by any horticultural increase, however; an autumn store of quinces is not easily acquired, but so rewarding a fruit is worth some effort.

As an *hors d'oeuvre*, quince sausages, from a German recipe of 1856, are best eaten after heating under a grill:

FRANKFURT QUINCE SAUSAGES

Peel quinces and reduce to a pulp by cooking and passing through a sieve; add an equal weight of sugar, dissolve sugar over a low heat, and stir together while heating until the mixture is stiff. Mix in chopped almonds, lemon peel, ground cinnamon and ground cloves. Stuff the whole into an artificial sausage skin and divide off into small sausages.

This Brazilian recipe is for quince soup; the German version, *Quittensuppe*, omits the rice and dilutes the cooked quince pulp with white wine before bringing to the boil:

SOPA DE MARMELOS

Peel and core a number of quinces, and cut into pieces. Cover with cold water. Add slices of lemon, a little sugar and a pinch of cinnamon. Bring to the boil and add a few tablespoons of rice. Cook until the rice and the fruit are tender, and pass through a sieve.

The quince goes well with a number of meats. Brisket of beef with quinces is a standard Romanian dish *(văcută cu gutui)*, and there are Persian meat-stuffed quinces *(dolmeh beh)*. This is an Israeli recipe:

QUINCES STUFFED WITH CHICKEN LIVER

Sauté 2 chicken livers in oil, remove, and sauté a chopped shallot in the remaining oil. Mince the chicken livers and shallot with 1 hard-boiled egg yolk, and mix in salt, pepper and 1 teaspoon tomato purée. This forms the stuffing. Next peel and core 4 quinces and boil in water until tender. Drain and sieve, and add 1 beaten egg, 3 tablespoons breadcrumbs, salt, pepper, and oregano. Knead this mixture into a dough, adding extra breadcrumbs if needed, and shape into squares; press a hollow in the centre of each and fill with the liver stuffing. Sauté the shapes lightly in oil, and serve hot.

Partridge and quince are both at their best in the autumn. There is a traditional recipe for combining the two – without the herbs that might be expected in a casserole because quince communicates its own overriding flavour:

CASSEROLE OF PARTRIDGE WITH QUINCE

Take 2 cleaned partridges, and fill with 1 oz (100 g) butter, well worked with salt, pepper and lemon juice. Place in a casserole lined with streaky bacon. Wrap over more slices of thin bacon. Core and quarter, but do not peel, sufficient quinces to fill out the casserole. Add 4 tablespoons Madeira and $\frac{1}{4}$ pint (150 ml) stock. Cook in a fairly hot oven (400°F/200°C/mark 6) for 45 minutes, add some lemon juice, and simmer for a further 5 minutes. The dish is now ready to serve.

The French *cotignac*, for which Orleans was famous and which Elizabethans called codinniack, and the Spanish *membrillo*, are old forms of quince paste that have endured in favour. The recipe for either is similar:

MEMBRILLO

Wash a number of quinces, leaving the peel intact. Quarter, and remove the seeds. Soften the quinces by steaming, and reduce to pulp by putting through a food mill. Add sugar to the pulp, weight for weight. Cook in a heavy pan until the paste begins to separate from the sides, then turn into shallow square tins and leave to cool. Ideally, the paste should dry in the sun for several days, but this process can be imitated in a warmed oven – not a lit one – in several sessions of an hour or two. The pink paste can be kept in the tins or in greaseproof paper, whole, or cut into sweetmeat squares for serving. Serve with coffee at the end of a meal.

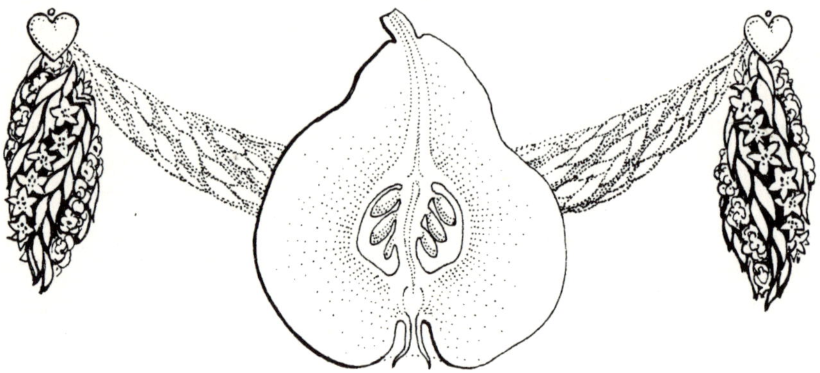

This mid-fifteenth-century recipe for a quince – or warden pear – dumpling is made with shortcrust pastry (see p. 11) in a pie dish or cake tin:

'QUYNCIS IN PAST'

'Take and make fair round coffins of fair paste; then take fair raw quinces, and pare them with a knife, and take fair out the core thereof; then take sugar enough, and a little powder ginger, and stop the hole full; and couch 2 or 3 wardens or quinces in a coffin and cover them and let them bake; and for default of sugar, take honey; but then put powdered pepper thereon, and ginger.'

The cover of the coffin (Greek *kophinos* = basket, frequently met with in cooking up to the eighteenth century) has to be a lid of pastry, brushed over with beaten egg; bake for about an hour in a moderately hot oven (375°F/190°C/mark 5), depending on the size of the quinces. Test to make sure the quinces are soft.

Quinces respond admirably to baking: unlike apples, they do not swell and burst, nor does the skin remain hard, but softens to a fine, glazed golden brown. Sir Isaac Newton's assistant wrote of him: 'In Winter Time he was a Lover of Apples, and sometimes at Night would eat a small roasted Quince.' His recipe would be:

ROASTED QUINCES

Brush and wash the quinces and core them, leaving them whole. Place them in a baking dish with a few spoonfuls of cold water. Stuff the centre of each with sugar, place a pat of butter on top of each and bake in a moderate oven (350°F/180°C/mark 4) for an hour. Serve hot, with cream.

This pudding is recommended by Dr Thornton in his *New Family Herbal* of 1810. Baking for half an hour in a moderate oven (350°F/180°C/mark 4) should suffice:

QUINCE PUDDING

'Scald your quinces tender, pare them thin, scrape off the pulp, mix with sugar very sweet, and add a little ginger and cinnamon. To a pint [600 ml] of cream you must put three or four yolks of eggs, and stir it into your quinces until they are of a good thickness. Butter your dish, pour it in, and bake it.'

This cold dessert comes from a Welsh cook's notebook of 1695:

MRS OWEN'S QUINCE FOOL

Quarter and core 1½ lb (675 g) of quinces; do not peel. Put these in a vegetable steamer that fits on the top of a saucepan, cover, and steam until quite soft. Sieve the quinces. Stir about ½ lb (225 g) castor sugar into the hot pulp and leave to cool. When quite cold, stir in $\frac{1}{3}$–$\frac{1}{2}$ pint (200–300 ml) fresh double cream, and serve in glasses.

NAMES: Anglo-Saxon *coin* or *quin*; Gk. *melon kydonion* (i.e. Cydonian apple); Latin *malum cotoneum* or *m. cydonium*; Fr. *coing*; Ger. *Quitte*; It. *cotogno*; Sp. *membrillo*; Port. *marmelo*; Russian *aiva*; Chinese *mu kua (C. sinensis)*; Japanese *marumeru (C. japonica)*.

STORAGE: Although apparently hard, quinces bruise easily and should be handled with care. Lay up in dry, cool conditions, away from other fruit, to which they would impart an alien flavour. The fruit should last the winter.

Raspberry

NATIVE to many European countries, the raspberry grows wild in Britain and elsewhere, as far north as the tip of Scandinavia. But it has also been cultivated since classical times, and was once abundant on Mount Ida in Asia Minor, inspiring Linnaeus to give it the name *Rubus Idaeus*. Adopted (in preference to the blackberry) as an English garden bramble, it is not mentioned as a domestic fruit before 1548. It has always been regarded as one of the best of dessert berries, simply to be taken with a little cream and sugar. Parkinson says, 'The berries are eaten in the Summertime, as an afternoones dish', and few writers recommend more than eating raw, perhaps with a touch of *fine champagne*. But cookery has some claims on the berry: one late seventeenth-century drink was raspberry sack, in which raspberry juice was added to grape wine, and jelly and jam were made from the fruit in the same period, as well as raspberry vinegar and raspberry brandy. The fragrant, cooling virtues of the berry are not lost in these recipes, though the heat tends to destroy some flavour: raspberries must not, on any account, be stewed. There are different varieties: the red summer-fruiting, the more delicate-tasting yellow (these are sometimes called 'white raspberries'), the black *(R. occidentalis)*, and a dark purple hybrid known as Black-Red Raspberries. The wineberry is related.

RASPBERRY SOUP

Simmer 2 tablespoons sago in 1 pint (600 ml) water with a stick of cinnamon until it turns transparent. Remove the cinnamon, and add 1–2 oz (25–50 g) washed, chopped and seeded raisins. Next add the juice of half a lemon, and finally the juice expressed from ½ lb (225 g) raspberries. Serve cold, with the addition of a few whole raspberries.

RASPBERRY SOUFFLÉ

Beat together, using a wooden spoon, ½ lb (225 g) ripe raspberries, 2 tablespoons cream, 2 oz (50 g) sugar and 2 oz (50 g) cornflour. Separate the yolks from 3 eggs, beat them, and then beat them into the raspberries. Add 2 oz (50 g) cake crumbs. Now whip the 3 egg whites to a stiff froth and fold into the mixture. Bake in a hot oven (425°F/220°C/mark 7) in a buttered soufflé dish for 25–30 minutes. Serve at once while still very hot.

RASPBERRY PUDDING (1)

Butter a pudding basin, and line it with pastry suet (see p. 13). Place in it half the raspberries that are to be used and half of a similar quantity of redcurrants. Squeeze over the fruit the juice of half a lemon, then coat it with sugar. Fill the dish with the remaining raspberries and redcurrants, again adding lemon juice and sugar, this time with 2 tablespoons cold water. Make a pastry lid, wet the edges, and place on top, joining the lining pastry and the lid. Press the edges and trim; cover with a floured cloth, buttered greaseproof paper, or aluminium foil and boil for 2½–3 hours.

RASPBERRY PUDDING (2)

Fill an oval ovenware dish with 1 lb (450 g) raspberries, sprinkle with a little sugar, and let them bake in the oven for 15 minutes, at a moderate heat (350°F/180°C/mark 4). Meanwhile, beat up ½ pint (300 ml) soured cream, 2 eggs, 1 tablespoon flour and 2 tablespoons sugar, and pour this mixture over the raspberries. Continue to bake at a slightly lower heat for about 25 minutes, until the top is golden brown. Sprinkle a little castor sugar on top and serve hot, with cream.

RASPBERRY FLAMBÉ

Melt 2 tablespoons butter in a small pan, and, keeping the heat low, stir in 1 teaspoon sugar until it turns a golden brown. Add 4 oz (100 g) fresh firm raspberries and stir gently but firmly to coat them with butter and sugar. Pour in 2 tablespoons raspberry brandy – *framboise* – and flambé. Let it simmer for a few minutes. Have a parfait glass containing some vanilla ice cream ready, pour the flambé sauce over it and serve with the addition of whipped cream and a sprinkling of a few almonds that have been roasted and coarsely ground. This serves one.

RASPBERRIES AND PORT WINE

Use 2 lb (1 kg) fresh ripe raspberries to ¼ pint (150 ml) port wine, and gently mix them together, working in ½ teaspoon finely grated lime peel and 1 tablespoon lime juice. After chilling for at least an hour, serve with whipped cream and biscuits.

RASPBERRY JELLY

Stone ½ lb (225 g) cherries and extract the kernels from the stones. Put this fruit, with 10 oz (275 g) raspberries, 10 oz (275 g) redcurrants and the kernels into a pan and add 1 pint (600 ml) water. Cook gently until soft and juicy, then pass through a sieve; to the juice that results add 2 oz (50 g) sago and a little sugar. Bring to the boil and simmer until the sago is cooked, then pour into a mould. When cold, serve with cream.

RASPBERRIES AND PINEAPPLE

Cut a ripe pineapple in half lengthwise, leaving the leaves but removing the core. Scoop the fruit out and cut into small pieces, placing it all in a basin with its own juice. Add raspberries to the same weight. Sprinkle the fruit liberally with castor sugar, chill in the refrigerator for an hour, then add 1 tablespoon kirsch. Fill one half shell of the pineapple with the fruit and the other half with stiff whipped cream that has been iced in the refrigerator. Serve the halves side by side.

RASPBERRY DELICE

Using any of the ways to make fools (see pp. 107–110), make a good bowl of raspberry fool, and chill it thoroughly. Serve it with ripe figs that have been peeled and cut in half and arranged round the edge of the bowl. The whole dessert should be very cold.

SOUFFLÉ DE RIZ AUX FRAMBOISES

Bring ¾ pint (450 ml) milk to the boil and sprinkle in 1½ oz (40 g) ground rice; stir in ½ oz (15 g) butter with a wooden spoon and cook for 5 minutes. Add the yolks of 2 eggs, mixed with 1 oz (25 g) sugar and 1 dessertspoon grated lemon rind, and then 4 oz (100 g) raspberries. Whisk the 2 egg whites until they are stiff, then fold into the fruit. Pour the whole into a buttered soufflé dish and bake for half an hour in a moderate oven (350°F/180°C/mark 4). Serve with raspberry syrup (see recipe for peach Melba, p. 203) or sauce (see p. 249).

NAMES: Rasbis, Hindberry (locally, Arnberry, Wood-rasp). Fr. *framboise*; Ger. *Himbeer*; Dutch *bramboss.*

STORAGE: Once picked, raspberries should be used immediately.

Red and White Currants

ONE of the first published aims of the Horticultural Society of London (now the Royal) was, in 1812, the production of 'a rich and sweet variety of the common *Red Currant*'. However, redcurrants remain little changed and rather too sharp for many tastes; the berries are only really acceptable if picked fully ripe in the autumn. Picked earlier, they are advisable only in making redcurrant jelly, perhaps the most widely known fruit jelly. Excellent as a separate accompaniment to game, to roast mutton in particular, and to hare and venison, redcurrant jelly was amongst the most popular of the new sauces to be introduced in the eighteenth century. For these meat dishes especially, it ousted the sauce commonly used since the Middle Ages, a concoction of red wine, cinnamon, sugar and breadcrumbs. For tarts and summer puddings, the redcurrant is excellent. It is a far more usual, and hardier, berry than the white currant, a dessert berry of delicate flavour that is rarely seen in the markets. This is in fact related to the redcurrant, simply lacking the pigment, but the red *(Ribes sativum)* is distinct from the blackcurrant (see p. 00), having a single main stem which can be trained against a wall. It is found wild, but it is generally cultivated in gardens in England. In Russia and Scandinavia the redcurrant is widely grown, and in France the fruit from around Bar-le-Duc provides a noted preserve of either red or white berries.

REDCURRANT JELLY

Wash a quantity of redcurrants, and strip them from their stalks. Bring them to the boil, with a little water, in a preserving pan, and let them simmer until a pulp is obtained. Strain this through a jelly bag, return the liquid to the pan, and for every pint (600 ml) add 1 lb (450 g) preserving sugar. Dissolve the sugar, then bring to the boil again; skim and boil for 3–5 minutes before testing to see whether the jelly will set quickly on a cold plate, and continue boiling until it does so. When ready, allow a short cooling period before pouring into warm jars and covering tightly with discs of waxed paper and cellophane.

MIXED CURRANT COMPOTE

Wash carefully 1 lb (450 g) stemmed redcurrants and ½ lb (225 g) each of white currants and blackcurrants. Put ½ lb (225 g) sugar and ¼ pint (150 ml) water, in a saucepan, dissolve the sugar over a low heat to make a syrup, and stew the redcurrants in this until they are soft, but not breaking up. Using separate saucepans, each with 4 oz (100 g) sugar and 5 tablespoons water to make a syrup, cook the two other fruits. When ready, arrange the fruits in glass bowls for each serving, keeping the fruits apart, half for the redcurrants, a quarter each for the white and black. Sprinkle with brandy or orange curaçao, and serve with thin cream.

REDCURRANT CREAM

Wash and prepare 1 lb (450 g) redcurrants and sprinkle with 6 oz (175 g) castor sugar. With ¾ pint (450 ml) milk and 2 tablespoons custard powder, make a custard and mix with this 6 oz (175 g) curd cheese (soft cream cheese, or fromage à la crème). Add the redcurrants to this, and decorate the top of the dessert with redcurrants. Serve chilled, with whipped cream.

DANISH FRUIT MOULD

Stew 1–1½ lb (450–675 g) redcurrants in a very little water, adding sugar to taste, and mash the fruit through a sieve. Measure the juice: to each pint (600 ml) take 1 oz (25 g) each ground sago and potato flour and make a smooth liquid of these with water. Boil the fruit juice, slowly add the thickening, and bring back to the boil, stirring constantly. Remove from the flame, and stir until cold. Thoroughly chill the fruit mould, and serve with whipped cream or ice cream.

NAMES: Fr. *groseille rouge, blanche*; Ger. *rote, weisse Johannisbeere*; It. *ribes rosso, bianco.*

STORAGE: When fully ripe, use within a few days. Do not chill or freeze.

Rhubarb

THERE are eight species of rhubarb, introduced into Europe from Asia (Mongolia, Tartary, China, the Himalayas and South Siberia are its homes; the roots used to be shipped from Canton, or sent overland from Russia) at various times from about 1608 *(Rheum rhaponticum)* to about 1828 *(R. emodi)*, when it first came to be used in both England and America as a culinary plant. It then began to appear in markets as a substitute for fruit, and it has been treated as a fruit ever since.

The rhubarb's green or red leaf-stalks, which like its roots and sometimes leaves were formerly confined to apothecaries' concoctions, soon became common fillings for tarts, pies and puddings. Much rhubarb was subsequently grown domestically for food in England and America ('pieplant' is an American synonym), where it flourishes as far north as Alaska, and in Germany, France, and particularly Switzerland. Rhubarb grows wild in the Central Asian provinces of Russia, where its astringency is favoured to such an extent that it is eaten raw as well as cooked.

The usual dishes include jams, sauces, fritters and dumplings, as well as pies and tarts, but rhubarb does not lend itself to much more than extensions of basically stewed stalks, and can be generally combined only with slightly acid fruits – segments of tangerine, orange and other citrus fruits, for instance, could be added to pies and puddings. For cooking, young stalks are best; old stalks thicken and are unpleasantly fibrous. Rhubarb is agreeably astringent to the taste, but invariably needs sugar; the purgative effect of the milder English, or Garden, rhubarb is well known. However, the leaves contain oxalates, and from time to time accounts of oxalic poisoning, sometimes fatal, are recorded; the temptation to boil or stew rhubarb leaves as if they were spinach should therefore be avoided.

RHUBARB TART

Prepare a shortcrust pastry lining for an open tart (see p. 11), partially cook (see p. 13) and set aside. Cut rhubarb up into inch ($2\frac{1}{2}$-cm) lengths, sweeten it with castor sugar, grate over it the rind of half a lemon, and cook in a very little water. When cool, drain and transfer to the pastry shell, and cook for a further 15 minutes in a moderately hot oven (400°F/200°C/mark 6). Beat $\frac{1}{4}$ pint (150 ml) cream into the remaining rhubarb juice, and whip until stiff, adding castor sugar as necessary. Pile this over the cooked tart when cool, garnishing with strips of angelica and glacé cherries. Serve quite cold.

RHUBARB AND STRAWBERRY TART

Cut up 1 lb (450 g) rhubarb into $\frac{1}{2}$-inch (1-cm) pieces, stew with sugar, and set aside, draining off the syrup in which it has been cooked. Make shortcrust pastry (see p. 11 – and include a dessertspoon of castor sugar) and roll into a thin round to fit a 9 or 10-inch ($22\frac{1}{2}$ or 25-cm) flan ring, press in and trim off the edges; after chilling for an hour, bake blind in a moderately hot oven (400°F/200°C/mark 6) for 10–15 minutes, until the pastry takes on a light colouring. Remove and let it cool. Now arrange the rhubarb pieces in rows in the shell, slightly mounded to the centre and with a little space round the edges for hulled strawberries in a ring – small whole strawberries or halved large ones. Add 2 tablespoons brandy to the cooking syrup, heat until it reduces to a thick liquid, then spoon it over the fruit. Cook for a further 10 minutes in a hot oven (425°F/220°C/mark 7), allow to cool before serving, and accompany with whipped cream.

RHUBARB CHARLOTTE

Cook 1 lb (450 g) young rhubarb, washed and cut into 1-inch (2-cm) lengths, over very low heat with 4 oz (100 g) sugar and a little grated orange peel. When the rhubarb is tender, leave it to cool. Melt 2 oz (50 g) butter over a low heat; remove from the heat and stir in first 2 oz (50 g) moist brown (or Demerara) sugar and then 6 oz (75 g) dry white breadcrumbs. Fry until golden brown. Arrange a layer of this mixture in a bowl and add a layer of the cooked rhubarb. Proceed with alternate layers of rhubarb and crumb mixture and finish with a layer of whipped cream. Serve cold.

NAMES: Fr. *rhubarbe*; Ger. *Rhabarber*; It. *rabarbaro*; Sp. *ruibarbo.*
STORAGE: Store in a cool and, if possible, damp place, bundling the stalks in newspaper; use within a few days.

Sauces

IF, as Brillat-Savarin says, the number of savours is infinite, and they are also differentiated by simple, double and multiple combinations, it is then impossible to classify them. This equally applies to the very large field of sauces, even to those made with fruit as a main ingredient. The word sauce derives from *salsa*, salt, and is applied to any soft or liquid preparation for dressing foods. Sauces work by contrast: ‘sweete meate will have soure sauce’ is an old maxim, and the Romans, on the evidence of Apicius, whose treatise on cookery has one whole book devoted to sauces, well knew the value of aromatics and seasonings, honey, vinegar and other ingredients to season meat and fish. The first fruit to enter into the sauces known in Roman Britain was the raisin; mediaeval England knew dried damsons and prunes as part of sauce recipes, and verjuice of grapes or crab-apples formed part of many early concoctions. Some of these mediaeval sauces are laboured attempts at sensation; 'gansel' or 'gauncil' was one, a goose sauce made by roasting a goose stuffed with quinces, pears, garlic, grapes and herbs, then removing the stuffing and combining it with wine and spices to make a pouring sauce. Bizarre sauces were a feature of Elizabeth I's reign, many of them using gooseberries, lemons or dried fruits among their many ingredients. They are too complex to recount; even an ordinary capon sauce combined claret, rose water, sliced oranges, ginger and cinnamon. Thinner sauces of gravy or wine appeared in the seventeenth century, and the classic sauces came with the rise of French cuisine in the eighteenth century. From then on, with a few exceptions, fruit sauces as such were conspicuous only in the regional cooking of countries such as Germany and the United States, and in Scandinavia and the Slavonic lands.

The distinction was now between sauces for meats, such as game or fowl, for which the astringent fruit element (such as cranberries) needed to be preserved almost fully, and sauces for puddings and ices, where softer fruits were more appropriate. These can always be simply improvised (on the model of the raspberry sauce noted here), often being little more than fruit in a jam-like state sharpened with brandy, liqueur or wine. While keeping a distinction from pickles and chutneys, there are many culinary possibilities in fruit sauces; the recipes below are an indication of the range.

The following is an excellent sauce for turkey:

ALMOND SAUCE (1)

Take 2 oz (50 g) blanched almonds and grind them finely. Melt about ½ oz (15 g) butter in a saucepan, and stir the almonds into this until they turn brown. Next sift in 1 heaped dessertspoon flour, add ¼ pint stock, and flavour with a little salt, a pinch of sugar and a touch of mace. Bring to the boil. Keep the sauce hot in a bain-marie, and stir in a tablespoon of cream before serving.

The Spanish way of making this sauce – *salsa de almendras* – is slightly different. This sauce is served with roast meats:

ALMOND SAUCE (2)

Roast 2 oz (50 g) blanched almonds in a frying pan, grind them, and work in enough butter to make a smooth paste. Transfer this to a saucepan with $\frac{1}{2}$ pint (600 ml) stock and simmer for half an hour. Have the yolk of 1 egg ready in a sauce boat, and pour in the sauce little by little, stirring it gently to bind with the egg.

This old-fashioned sauce is intended specifically for Christmas puddings:

MARIA HUDSON'S ALMOND SAUCE

Blanch 6 bitter almonds and 6 sweet almonds and pound them well. Cream 4 oz (100 g) unsalted butter with 4 oz (100 g) castor sugar. Add the pounded almonds, and mix thoroughly. Then add 4 tablespoons brandy, drop by drop. Set the sauce aside for 12 hours before using; it should be quite stiff.

Apple sauce goes well with goose, roast pork and other pork dishes:

APPLE SAUCE (1)

Peel and core 3 cooking apples, and slice them. Place them in a saucepan with 6 sautéed almonds, a chopped onion, the grated rind of an orange and a slice of toasted bread dipped in vinegar. Add pepper and mixed spices to taste, moisten with white wine and vinegar in equal quantities, and boil until soft. Pass the sauce through a sieve, and serve hot.

APPLE SAUCE (2)

Pare, core and slice 1 lb (450 g) cooking apples. Place in a saucepan with a minimum of water over a slow flame; cover, and leave until the apples form a pulp. Drop in a lump of butter and sugar to taste, and beat well. Serve hot.

There are several variants of this sauce for cold meats, some incorporating chopped hard-boiled egg:

APPLE AND HORSERADISH SAUCE

Peel, core and dice 1 lb (450 g) cooking apples and place them in a saucepan with $\frac{1}{4}$ pint (150 ml) dry white wine. Bring to the boil over a moderate heat, simmer for 10 minutes, and continue to cook for as long as it takes to reduce the liquid by half. Then make a purée of the fruit, let it cool in a bowl, and stir in 2 teaspoons mayonnaise, followed by 2 teaspoons freshly grated horseradish. Add salt and pepper to taste. Serve cold.

APRICOT SAUCE

Mix together 2 tablespoons apricot jam, 2 oz (50 g) sugar and 4 tablespoons white wine. Heat them thoroughly for 5 minutes.

This is the standard Mexican sauce, served with meat dishes, seafood, *tortillas* and other dishes. Lime juice, salt and French dressing are sometimes added to make a salad dressing:

GUACAMOLE (AVOCADO SAUCE)

Mix together the flesh of 3 large avocados, chopped, an onion and a medium-sized tomato, peeled and chopped, 2 small finely chopped and de-seeded chillies, 1 tablespoon olive oil, 1 teaspoon vinegar and 2 teaspoons salt. When the guacamole turns creamy, it is ready.

CHERRY SAUCE

Stone $\frac{1}{2}$ lb (225 g) cherries, break the stones and add the kernels to the fruit. Pound the kernels and the fruit, then put them in a small saucepan with $\frac{1}{4}$ pint (150 ml) of red wine and the same amount of water. Add to this a spoonful of sugar, a pinch of ground cinnamon and a piece of lemon rind. Simmer until the cherries are cooked. Put them through a sieve and return to the saucepan, stirring in a little arrowroot that has been mixed to a smooth paste with cold water. Bring back to the boil. Add a few fresh, stoned cherries and serve.

Chestnuts may be simply boiled until soft, peeled, rubbed through a sieve, and heated with a little gravy and sherry; this will give an adequate sauce for chicken or turkey. A more complex sauce, for venison or hare, is as follows:

CHESTNUT SAUCE (1)

Fry 1 large onion in $1\frac{1}{2}$ oz (40 g) butter until brown and then add 1 oz (25 g) of flour. Let this brown, then add $\frac{3}{4}$ pint (450 ml) of stock. Add a bayleaf, a pinch of sage, a pinch of thyme, 6 peppercorns and salt to taste. Simmer for 20 minutes, covered, and strain. Now add 12 chestnuts that have been boiled and sieved into a purée, work in 1 tablespoon butter and stir in 5 tablespoons port. Serve hot.

This one is for turkey, or mutton cutlets:

CHESTNUT SAUCE (2)

Prick 20 shelled chestnuts and boil them, then peel them and pound into a smooth paste. Put this in a saucepan with 1 oz (25 g) butter, 1 oz (25 g) chopped mushrooms, and pepper and salt as necessary; stir and heat. Serve as hot as is acceptable.

This is a simple sauce to serve with meat and poultry, and the usual American sauce served with turkey:

CRANBERRY SAUCE

Wash 12 oz (350 g) cranberries and cook in ¾ pint (450 ml) water for 10 minutes only with 12 oz (350 g) sugar. Cool, and skim as necessary.

This sauce is traditionally one of those for sucking pig, a dish few households would now tackle, but it could be used for most pork dishes, or, if the wine is changed from white to red, for hare or venison:

CURRANT SAUCE

Into ½ pint (300 ml) boiling water put 1 oz (25 g) currants, 2 cloves, a little grated nutmeg, 1 oz (25 g) fresh breadcrumbs, 4 tablespoons sweet white wine, sugar to taste, and a small piece of butter. Simmer these for a few minutes, and serve hot in a sauce boat.

This is an Arab sauce to accompany *chilau*, a dish of steamed rice cooked by a Persian method:

DATE AND ALMOND SAUCE

Wash 1 lb (450 g) basmati rice, soak for a few hours in salt water, drain, rinse, and cook until nearly tender in boiling salted water. Drain the rice, and keep warm. Prepare the sauce from 4 oz (100 g) blanched almonds, halved and fried in 2 oz (50 g) butter until golden, add 2 oz (50 g) raisins and 4 oz (100 g) fresh dates, chopped into small pieces. Stir the mixture over a moderate heat for a few minutes, add ¼ pint (150 ml) water and simmer very gently until the water has been taken up by the fruit. Take a large saucepan and melt in it 1 tablespoon butter, add half the rice over it, then the date and almond sauce in an even layer, and lastly the remaining rice. Cover with a cloth and a lid. Steam over a low heat for half an hour.

GOOSEBERRY SAUCE

Put ½ lb (225 g) green gooseberries in a saucepan with a little water. Cook them slowly. When they are soft, drain off any excess liquid, rub the fruit through a strainer, and return the purée to the saucepan. Add 2 tablespoons sugar, ¼ pint (150 ml) single cream, ½ oz (15 g) butter and some grated nutmeg. Heat, stirring continuously but do not boil. Serve hot.

The following is the standard sauce for grilled mackerel:

GREEN GOOSEBERRY SAUCE

Boil ½ lb (225 g) green gooseberries. Drain off the water, and pass them through a sieve. Put the purée into a saucepan and add either 4 tablespoons spinach juice or some raw green sorrel. Add a piece of butter, and a seasoning of nutmeg, sugar, salt and pepper, and heat. The sauce is best served very hot.

This plain German lemon sauce, *Zitronensauce*, is to go with puddings:

LEMON SAUCE

Beat the yolks of 2 eggs with ½ pint (300 ml) boiling milk. Add the grated rind of half a lemon, a little sugar and a pinch of salt. Thicken the sauce with 2 teaspoons cornflour, mixed with a little cold milk, stir well, strain, and serve hot.

LEMON SAUCE WITH CREAM

Into a saucepan containing 2 tablespoons white wine and the same quantity of water, 3 egg yolks and 2½ tablespoons sugar, squeeze the juice of a lemon and grate the zest. Whisk energetically over the heat until the sauce thickens. It must not boil. Remove from the heat and continue to whisk until cold. Stir in ½ pint (300 ml) cream that has been partially whipped.

The following is the basic lemon sauce of Greek cookery. It goes well with fish, chicken, white meat and stuffed vineleaves:

AVGOLEMONO SAUCE

Separate the yolks and whites of 3 eggs. Beat the whites with a pinch of salt until they are foamy but not stiff. Add 1 egg yolk at a time, and beat until fully blended, then gradually beat in the juice of 2 lemons and ½ pint (300 ml) boiling stock, slowly so that the eggs do not curdle. Cook until the sauce thickens, stirring rapidly and not letting the mixture boil.

This sauce goes with fish or chicken:

OLIVE AND ALMOND SAUCE

Make ½ pint (300 ml) Béchamel sauce from milk, white roux, onion, peppercorns and herbs in the usual way. Add to this 5 tablespoons blanched, shredded almonds, 8 green olives, stoned and cut in quarters, 1½ teaspoons lemon juice and a sprinkling of cayenne pepper.

Edward Spencer Mott's (Nathaniel Gubbins') sauce for cold mutton:

ORANGE SAUCE

Grate the thin yellow rind of 2 oranges and put them in a bowl. Pour 4 tablespoons port over it and ¼ pint (150 ml) dissolved redcurrant jelly, 1 teaspoon mixed mustard, a finely chopped shallot, a pinch of cayenne pepper, and finally some more thinly pared orange rind. Add sugar to taste. Mix well, heat up, strain, and bottle.

This hot orange sauce is for wild duck or game:

HOT ORANGE SAUCE

Cut the thinly pared rind of an orange into long strips. Pour boiling water on it, cook for 5 minutes and drain. Now add to the peel the juice of 2 oranges and 1 lemon and ½ pint (300 ml) rich brown gravy made from brown roux and stock. Add pepper and salt, stir and heat. Serve hot.

One of a number of versions of this established sauce for duck or duckling, this could be fortified with a few drops of curaçao added as a final touch:

SAUCE BIGARADE

Cook 1 oz (25 g) sugar with 1 tablespoon wine vinegar until pale caramel, add ½ pint (300 ml) stock and use this to dilute the juices from cooking the duck or game. Boil rapidly for about 5 minutes. Add the juice of 1 orange, a squeeze of lemon juice and 2 tablespoons julienne strips of blanched orange rind.

This German sauce is for roast poultry, to be served separately:

PETERSBURGER OBSTSAUCE

Peel, core and grate 1 apple, peel an orange and chop finely. Add these, with the juice of 2 lemons and ¼ pint (150 ml) stock, to ½ pint (300 ml) white wine which has been heated. Serve hot.

This Persian sauce *(khoresh-e-fesenjan)* accompanies chicken breasts which have been halved and poached in a beef broth:

POMEGRANATE AND WALNUT SAUCE

Take 5 oz (150 g) ground walnuts and ¼ pint (150 ml) squeezed pomegranate juice (see p. 221) and mix in a saucepan with 1 tablespoon sugar. Bring to the boil and then simmer over a low flame for 15 minutes. Add this to ½ pint (300 ml) of the poaching liquid of beef broth, to which an onion sautéed in butter has been added. Add cinnamon, nutmeg and pepper, bring to the boil again, and simmer for up to 30 minutes, until the sauce turns thick and dark. Add lemon juice before serving.

PRUNE SAUCE

Soak ½ lb (225 g) stoned prunes in 1½ pints (900 ml) water overnight. Use the same water to bring them to the boil, and add 2 tablespoons sugar. Continue cooking until the prunes disintegrate, then pass through a sieve. Reduce or add more water as necessary, and serve hot.

A seventeenth-century sauce for veal, which was sometimes served with a cucumber sauce, a mint sauce, or a whole orange. The carrot water may be replaced with tap water:

QUINCE AND CRAB-APPLE SAUCE

Strain off the water in which carrots have been cooked, and flavour it with quinces cooked to a pulp and the juice of some crab-apples. To this add a little claret, some sugar, the juice of a lemon, a pinch of nutmeg, pepper and salt, some 'rose vinegar' (use ordinary vinegar) and a little ground clove; reduce as necessary and serve hot.

A Dutch sauce, *rozijnen saus*, is particularly good with smoked ox tongue:

RAISIN SAUCE

Dissolve 4 oz (100 g) brown sugar in $\frac{1}{2}$ pint (300 ml) stock from cooking the tongue, add 3 oz (75 g) raisins and the juice of half a lemon and simmer for 10 minutes. Add $\frac{1}{4}$ pint (150 ml) white wine and 1 teaspoon grated lemon rind. Stir in 2 tablespoons butter. Serve the sauce separately.

RASPBERRY SAUCE

Wash 1 lb (450 g) fresh raspberries and pass them through a sieve, then combine 4 oz (100 g) castor sugar with the purée in a blender and stir in 2 tablespoons kirsch. Chill thoroughly before serving.

This sauce goes with steamed puddings:

REDCURRANT SAUCE

Put 2 oz (50 g) lump sugar in a saucepan, cover with water, dissolve the sugar over a low heat, and boil gently; strain the syrup into a saucepan and add 1 tablespoon redcurrant jelly and the grated rind of half a lemon and half an orange. Bring back to the boil, and simmer until the jelly has melted. Stir constantly. Strain, and add 1 tablespoon maraschino. Serve hot.

A Russian sauce for baked fish, such as carp or bream, or for boiled tongue. As a sauce for fish, a fish stock would be preferable to the brown sauce:

SULTANA AND PRUNE SAUCE

To a basic brown roux, using ½ oz (15 g) butter and ½ oz (15 g) flour, add 2½ teaspoons sugar, half a lemon, thinly sliced, 2 oz (50 g) sultanas, washed and scalded in boiling water, 1 oz (25 g) chopped almonds, 2 oz (50 g) prunes, stewed and stoned, a little of their juice, and 4 tablespoons white wine. Bring to the boil and serve hot.

This is adapted from a mediaeval recipe for 'stockfysshe' with a sauce of ground cloves of garlic, pepper and walnut kernels, the stockfish being in all probability ling rather than dried cod:

LING WITH WALNUT SAUCE

Soak a dried ling and boil twice, renewing the water and simmering the second time in a shallow pan. Shred onions over the fish, cover, and let them cook until tender. The sauce is made with butter, flour, pepper and chopped pickled walnuts, to which are added the onions from the fish and some of the fish broth; boil this up. Drain the fish, pour the sauce over it, sharpened with a little vinegar, and garnish with watercress and a few halved pickled walnuts.

This Russian sauce is for trout, *sousye 'satsyebyeli' (oreyekhovyi sous)*:

WALNUT SAUCE (1)

Mix ½ lb (225 g) finely chopped walnuts with ½ oz (15 g) pounded garlic and a pinch each of salt and cayenne pepper. Pass the mixture through a mincer. Then add 2 tablespoons chopped mint, 8 tablespoons wine vinegar and sufficient clear fish stock to form a sauce. It is served separately from the fish, which should be cooked in the ordinary way in water to which wine vinegar has been added. Apart from the sauce, a garnish of radishes, spring onions and chopped parsley accompanies the trout.

An eighteenth-century recipe for a sauce for game and baked fish:

WALNUT SAUCE (2)

Take ½ pint (300 ml) of the juice of pickled black walnuts and add 2 of the walnuts, chopped, 6 chopped capers, 2 teaspoons anchovy paste, 2 teaspoons finely chopped chives and 1 teaspoon brandy. Simmer, then rub the whole mixture through a sieve.

Strawberry

THE berry that the Elizabethan physician William Butler supposed God had not surpassed was the wood strawberry, probably wild, and rarely tasted nowadays. Until Thomas Laxton began an extensive programme of breeding strawberries in the 1860s, following the pioneer work of T. A. Knight, the consumption of the fruit was limited to varieties quite unlike those normally eaten today. The history of the strawberry *(Fragaria)* goes back to the Romans, but there are few ancient references to it; Pliny is in fact the last known classical writer to refer to the berry as among the natural products of Italy, and there is then a gap of some 1,200 years before the strawberry is noted in the works of a Greek physician. The strawberry was certainly in cultivation in the fourteenth century; references to it abound, and the plant and its berries appear in many illuminated miniatures and religious paintings – a symbol of righteousness and also, as in Bosch's 'Garden of Delights', where a naked group balances a giant berry, of voluptuous living.

It is often asserted that Cardinal Wolsey introduced the dish of strawberries and cream at a banquet he gave; certainly by later Tudor and Stuart times the berry was much in demand. It merited many ornate silver strawberry dishes – the berry abhors base metal – and the opening up of vast strawberry fields in Brentford, Isleworth and Twickenham, whence strawberry women carried head-loads to Covent Garden. Here, by 1849, Mayhew estimated that 600,000 pottles of strawberries – with several pounds of fruit in each – were sold a year, such was people's enduring and insatiable delight in the fruit. The demand continues – though for very different berries to Butler's; breeders are now causing them to approach Bosch's in size (a Belgian variety Souvenir de Charles Machiroux is huge, and the Japanese Fukuba variety may reach 3 inches ($7\frac{1}{2}$ cm) long). Among the steady and persistent efforts of horticulturalists to improve and diversify the strawberry – one of the best documented of fruits – it may be valuable to distinguish five classes:

1 The Alpine, or Perpetual, or Wood Strawberry *(F. vesca)*. This is the French *fraise des bois* or *fraise des quatre saisons*. It is the species that was generally gathered in Europe before the Virginian Strawberry was introduced.
2 The Hautbois Strawberry *(F. elatior)*, called *capron* by the French; musky in flavour, it has about eight varieties.
3 The Green Strawberry *(F. collina)*, a green, red-tinged berry of Europe and northern Asia, with a musky pineapple flavour.
4 The Scarlet, or Virginian, Strawberry *(F. virginiana)*, whose home is New England. It was first noticed in Massachusetts in 1621, and seems to have grown in abundance. The Indians crushed it with meal to make bread. Parkinson mentions it as being in England in 1629, but its arrival is not documented, and it was a hundred years before serious attempts to improve it began.
5 The Pine or Common Garden Strawberry *(F. chiloensis)*. The most important of strawberries, whose origin lay in the journey of the army engineer Amadée Frezier to Chile on a reconnaissance mission for Louis XIV. Frezier (curiously, his name derives from the French for strawberry and was shared with that part of the

family who emigrated to Scotland as Frazers) reached Concepcion in 1712, and in the course of sketching fortifications, the physical geography of the region and any new plants he noticed, he described an exceptionally large-fruited strawberry he found in a valley outside the city. He brought back plants to France, and five survived the journey to Marseilles. The subsequent story of the complex genetics of the strawberry, of the involvement of Linnaeus and the great French naturalist Antoine Duchesne, is a long one, but it is no exaggeration to say that one of Frezier's plants is the ancestor of the modern strawberry.

So general is the habit of eating strawberries out of hand or with the addition of cream, in degrees of sophistication from the 'Eton mess' to *fraises des bois* with white wine and thin cream, that the fruit features in few traditional recipes. Cooking certainly spoils the strawberry. Strawberry jam only just allows the berry to survive with an identity, and tinned strawberries are unspeakable. The conclusion must be that as little as possible should be done in the kitchen to this best of berries.

STRAWBERRY SOUP

Mash 1 lb (450 g) strawberries, setting aside 6 of the best berries. Sprinkle sugar on the purée, as needed, and add $\frac{1}{4}$ pint (150 ml) water. Leave for 2 hours, then pass through a wire sieve. Add $\frac{1}{2}$ pint (300 ml) claret, $\frac{1}{2}$ pint (300 ml) water, and the juice of a quartered lemon. Refrigerate for 3 hours. After adding the whole strawberries, serve this Swedish *fruktsoppa* with almond paste balls (see p. 16).

STRAWBERRIES AND CREAM

Select the best wild strawberries, and hull and wash them with care. Place in a china bowl, sprinkle with the best quality kirsch, and serve with thin cream.

STRAWBERRIES AND ORANGE

Separate the smallest and best strawberries from the quantity to be used and set them aside. Pass the others through a sieve to make a purée and mix into this the juice of 2 oranges. Sweeten as necessary, and decorate the purée with the better strawberries.

STRAWBERRIES AND BRANDY

Crush as many strawberries as are needed in Sauternes and add a dash of grated lemon. Mix into this a half wineglass of brandy, preferably a *marc* (and ideally a Cordial Medoc, the strawberry-flavoured *marc*), and cover the top with whipped cream sweetened with castor sugar. Chill, and sprinkle over chopped almonds.

FRAISES CARDINAL

Wash and hull 2 lb (1 kg) fresh strawberries and arrange in glass dishes, one for each serving. Prepare a raspberry sauce by simmering, for about 2 minutes, 5 tablespoons raspberry jam or conserve, 2 tablespoons sugar and 5 tablespoons water in a saucepan, adding last of all 1 tablespoon kirsch. Chill the sauce, and pour it over the strawberries; sprinkle each glass with slivered almonds and slivered pistachios.

STRAWBERRIES IN JELLY

Wash and hull 1 lb (450 g) strawberries of equal size and set them on one side to soak in sugar and brandy. Make about 2 pints (1·2 litres) claret jelly, using 2 oz (50 g) gelatine to ½ pint (300 ml) water, 1 bottle of claret and a small pot of redcurrant jelly; stir over a low heat until the jelly is a fine, clear red. When the jelly is nearly cold, set some of it in a wetted jelly mould and leave it to set. Arrange a layer of strawberries over this, then pour more jelly over. Let this set, and repeat the layers as many times as necessary to fill the mould. Chill thoroughly before turning out.

STRAWBERRY PUDDING

Line a 2-pint (1·2-litre) mould with lemon jelly – setting some jelly aside – sprinkle it thickly with blanched and chopped pistachio nuts, and let it set in the refrigerator. Rub approximately 1 lb (450 g) strawberries through a sieve, then add a little castor sugar and ½ oz (15 g) dissolved gelatine to each ½ pint (300 ml) of sieved strawberries. Chill, and when on the point of setting, blend this mixture with ½ pint (300 ml) thick cream and whisk until it stands up in points. Separately, dissolve ½ oz (15 g) gelatine in ¼ pint (150 ml) warm water, and when it is cold add ¼ pint (150 ml) milk and sugar to sweeten. Then blend in ½ pint (300 ml) double cream and flavour with vanilla, whisking until it is the same consistency as the strawberry cream. Now fill the mould with alternate layers of the two creams, and cover with some more of the lemon jelly. Let the pudding chill thoroughly in a refrigerator, and turn out on a dish before serving.

STRAWBERRIES À LA MUSCAT

Hull a quantity of strawberries and marinade them in a syrup made as follows: add 1 teaspoon lemon juice to ½ pint (300 ml) water, dissolve 1½ lb (675 g) white sugar in it over a low heat and bring to the boil. While it is boiling throw in 2 handfuls of fresh elderflowers. Let these boil for 3 or 4 minutes in the syrup, then remove them. Serve the strawberries and syrup, which will taste of muscat, with cream cheese that has been beaten to a smooth purée.

ERDBEERSCHNITTEN

Fry in butter, on one side only, slices of fresh white bread until they are crisp. While hot, completely cover the uncooked side with strawberries. Sprinkle with castor sugar, and add thick cream.

SOUFFLÉ AUX FRAISES DES BOIS

To prepare a soufflé mixture place in a saucepan 2 oz (50 g) flour, $\frac{1}{2}$ lb (225 g) sugar and the yolks of 5 eggs; mix well over the heat by adding, little by little, 1 pint (600 ml) milk that has been brought just to the boil. Cook, and whip until the mixture is like a thick cream, then let it get cold. Now add 6 egg whites beaten to a stiff froth, and $\frac{1}{2}$ pint (300 ml) purée of wood strawberries, mashed with a little castor sugar and whipped cream until very smooth. Lightly butter a soufflé dish, sprinkle it with sugar, and put the batter in the dish to about three-quarters level. Cook in a moderate oven (350°F / 180°C / mark 4) for 14–16 minutes. Serve immediately.

This recipe comes from the Regency cookery book *The Housekeeper's Instructor*, published in 1811:

STRAWBERRY FRITTERS

'Make a batter with flour, a spoonful of sweet oil [use almond oil], another of white wine, a little rasped [grated] lemon peel, and the whites of two or three eggs; make it pretty soft, just fit to drop with a spoon. Mix some large strawberries with it, and drop them with a spoon into the hot frying pan. When of a good colour take them out, and drain them on a sieve. When done, strew some sugar over them, or glaze them [sprinkle sugar on them and melt under a very hot grill] and send them to table.'

OMELETTE SUZETTE

Have ready a quantity of strawberries that have been soaked in kirsch for an hour or two. Make a few small omelettes, taking care not to overcook them, and stuff them with the strawberries: roll the omelettes, skewer them with a wooden skewer, dip them in a fritter batter (see recipe for Bundaberg banana fritters, p. 44), and fry them all over in hot fat. When fried, drain, and sprinkle with castor sugar, if possible with vanilla sugar (where a vanilla pod has been kept in a castor sugar jar).

NAMES: Fr. *fraise*; Ger. *Erdbeer*; It. *fragola*; Sp. *fresa*.
STORAGE: Ripe strawberries keep a few days at most in a dry cool place; refrigerating affects their moisture content and taste.

Tangerine and other Citrus Fruits

THE orange, the lemon, the grapefruit, and the lime are the principal members of the genus *Citrus*. There are many others however, some of them hybrids, not of as high a nutritional value, but useful substitutes where a slight change of taste is sought.

Bergamot *C. bergamia*, an orange extensively cultivated in Calabria for its essential oil and flavouring rind.

Calamondrin Orange *C. mitis*, the Panama orange: an ornamental tree, but the rind of the fruit is edible and used in marmalade, and the juice for seasoning.

Citrange A sweet orange crossed with a trifoliate orange.

Clementine A cultivated variety of the tangerine, as is the seedless satsuma (Japanese, named after the province, but widely grown in Spain).

Kumquat *C. japonica*, but Chinese in origin. This is a miniature golden orange, delicious when preserved, pickled or candied as an entire fruit.

Naartje The native South African tangerine, the country's best citrus fruit.

Ortanique A superior orange-tangerine hybrid, the fruit is especially succulent but difficult to disengage from its skin; a West Indian favourite.

Tangerine Synonym Mandarin (French *mandarine*); the loose-skinned *C. reticulata* from China, an orange with the advantage of small size, sweet juice, and an easily removed peel. Good in salads and in most dishes where oranges are normally advised.

Ugli The trade name of what in N. America is known as the tangelo and what the West Indians call 'hoogly'; a yellow-green, rough-skinned hybrid of the grapefruit and the mandarin, shapeless, with a deep inside pith and some loose reticular strands. It has a most refreshing juice.

Wampi An admired fruit of Thailand *(Clausena excavata* Burm. f.*)* with yellow, marble-sized fruit, with a flavour somewhere between a grape and a gooseberry.

TANGERINE SALAD

Peel 4 large tangerines, remove the pith, slice the fruit and remove the pips. Arrange these on a dish, and shape round them a ring of whipped cream – ¼ pint (150 ml) will be needed. Cut up 2 ripe, peeled pears, sprinkle them with lemon juice to prevent discoloration, and arrange the pieces over the tangerines. Finally, squeeze the rind of the tangerines over the pears, and decorate with 2 oz (50 g) finely chopped preserved ginger. Serve very cold.

TANGERINE CREAM

Extract the juices from a number of tangerines, and strain to provide 1½ pints (900 ml). Add ½ pint (300 ml) water, and 6 tablespoons granulated sugar, and bring to the boil in a saucepan. Stir into this 4 heaped tablespoons custard powder, diluted with an equal amount of water. Cook over a low flame for a few minutes, stirring the mixture. Take the pan off the heat and fold in 1 teaspoon finely grated tangerine rind. Separate an egg; whip the egg white, then add the yolk, and fold the combination into the custard. Let the whole cream grow cold, and serve in small glasses. Decorate each serving with a number of tangerine segments.

STORAGE: Like oranges and lemons, citrus fruits are among the most durable of fruits. Eventually the skin hardens and the fruit dehydrates, or patches of mould appear, or simply patches of softening. These can be sliced off if the rest of the fruit is healthy.

Tropical and Sub-Tropical Fruits

SOME regions of the world, notably the Caribbean, Central America and the Malaysian Archipelago, have a prodigality of lesser known trees and shrubs that bear edible fruit. Some of these are native, others – particularly in the West Indies – come from the Old World. Few are cultivated, and few are cooked. Of necessity, most travel no further than local markets, to be eaten while briefly at their best, but a number of fruits, mainly from the Caribbean, now reach British and other temperate zone markets.

Akee *(Blighia sapida)* Brought to Jamaica by Captain Bligh from West Africa, this peach-sized fruit turns yellow and red when ripe and splits open to display three large black seeds in three segments of white edible flesh. This fruit, poisonous when unripe or over-ripe, can be stewed or browned in butter; it is eaten with fish and rice in the Caribbean.

Barbados Cherry *(Malpighia glabra)* A garden fruit tree of Jamaica and Barbados, with bright red lobed berries the size of a cherry, sweetish, with much juicy pulp and triangular, fluted seeds. Its taste is more like the raspberry, and it is used for jam, jelly, stewed or preserved fruit, and pies.

Biriba *(Rollinia pulchrinervia)* A Brazilian tree fruit, among the best of its kind, related to the soursop and sugar apple. It is pale green or yellowish, with thick, soft, spiky projections, with cream-coloured juicy flesh of a pleasant flavour; it is generally known as *fruta de condessa*.

Cape Gooseberry *(Physalis peruviana)* One of the more important cultivated fruits of the Cape of Good Hope, although South American in origin. The roundish fruit is yellow and enclosed in an inflated 'Chinese lantern' calyx. Grown in India, Australia and the south of France, the fruit makes excellent jams and jellies.

Ceriman *(Monstera deliciosa)* A luxuriant climbing plant with large incised and perforated leaves of shining dark green; it has long cone-like fruits with a banana-and-pineapple taste, pleasant when fully ripe, but with irritating crystals of oxalate when not so. Tropical American, it is grown in many greenhouses in temperate countries.

Chayote *(Sechium edule)* Related to the squash and melon, this pear-like fruit has spread from Mexico to the West Indies and Australia. It is a vine, and the white or greenish hanging fruits are abundant. The crisp flesh can be boiled, fried or sliced for salads – it serves both as vegetable and fruit.

Cherimoya *(Annona cherimola)* A native of Peru, a small tree of the custard apple family (see below) with a large heart-shaped fruit of 4–14 lb (2–7 kg) in weight. The flesh is creamy, resembling the soursop, and generally admitted to be the best of the Annonaceae in flavour and texture.

Coconut *(Cocos nucifera)* The familiar palm, found almost everywhere in the tropics, bears fruit continuously. A source of cooking oil, coconut milk – much used in tropical dishes – and crisp coconut for grating and use in cakes, pies and puddings and for flavouring.

Custard Apple *(Annona reticulata)* Central American, known as 'Bullock's Heart' in the West Indies. When completely ripe, the skin of the fruit blackens and the creamy flesh becomes sweet and aromatic, and soft enough to eat with a spoon.

Durian *(Durio zibethinus)* A fruit tree of the Indian archipelago, with globose, pendulous fruits, a thick rind covered with coarse pyramidal spines, four-celled internally, with a buttery aromatic flesh. Combining a delicious flavour with an exceptionally repulsive odour, the durian is prized by connoisseurs, but its use is restricted by law in Malaya.

Jackfruit *(Artocarpus heterophylla)* An Indian tree, introduced into the American tropics. The water melon-sized fruits have prickly rind and brown seeds that are edible when roasted; the yellow pulp about them can be boiled or cooked in various ways.

Jujube (tsao) *(Zizyphys jujuba)* An East Indies fruit, introduced into China about 1,300 years ago, where it has become an important fruit. Reddish brown at maturity, the fruit is the size of an olive and has sweet whitish pulp.

Langsat *(Lansium domesticum)* Small velvety fruits in grape-like clusters grow on this tall tree, known chiefly in Malaysia. A thick skin encloses a few segments of white translucent pulp, subacid, juicy and aromatic.

Loquat *(Eriobotrya japonica)* A Japanese and Chinese tree, cultivated in the south of France, in America (as the Japanese plum) and elsewhere. The soft yellow fruit is uniquely flavoured, refreshing, sweet and acid, reminiscent of apples and pears.

Mammee *(Mammea americana)* A tall tree with dense green foliage and large globose brown fruits 'like rusty cannonballs'; the flesh can be cooked, the flavour being described as between apricot and raspberry.

Mangosteen *(Garcinia mangostana)* This Malayan fruit is claimed to be the most delicious of tropical fruits. About the size of a crab-apple, the rind is thick, dark red to purplish outside and pale violet inside, and contains six snow-white arils of exquisite flavour. It has not succeeded outside Malaya.

Otaheite Gooseberry *(Phyllanthus acidus)* Deriving its name from Tahiti, this tree was introduced into the West Indies in 1793, beautifully ornamental, with small light yellow fruit hanging in countless clusters. Tart and gooseberry-like in flavour, the fruit turns red when cooked with sugar; a good jelly is made from them.

Prickly Pear *(Opuntia tuna,* etc.*)* All types of spiny cactus produce flowers which develop edible fruits, generally red or yellow, and with an agreeable pulp, but detracted from either by excessive seeds or by cushions of minute spines.

Rambutan *(Nephelium lappaceum)* A fruit of tropical Asia and Malaya, this is a larger form of the lichee and has similar characteristics. The shell is brittle and has long, soft spines. It is much esteemed.

Rose Apple *(Syzygium jambos)* Found in eastern Asia and the West Indies, this graceful, ornamental tree is sweet-scented with fruit the size of small apples and a crisp, rose-like fragrance.

Sapodilla *(Achras sapota)* This South American tree has a round, brown berry with a sweet pulp, both eaten out of hand and made into syrups and jams. It is known as the Naseberry in the West Indies.

Sapote *(Calocarpum mammosum)* A Central American tree with a large fruit, like the mammee, oval and brown-skinned with firm, spicy, reddish flesh, either eaten fresh with sugar or made into a conserve.

Star Apple *(Chrysophyllum cainito)* Long cultivated in the West Indies, and a common garden tree, some trees bear white and some purple fruit. When cut, these reveal eight star-like segments. The flesh is melting and sweet, and of pleasant flavour, usually eaten fresh.

Surinam Cherry *(Eugenia uniflora)* This prized ornamental shrub has red leaves and eight-ribbed, red, glossy fruits, thin-skinned and juicy with a spicy, subacid flavour. The cherries are used as dessert fruit, stewed, or made into jams and jellies.

Tamarind *(Tamarindus indica)* A native of East Africa, the large tamarind tree has fruits in the form of brown curved pods, with seeds embedded in a sweet-acid pulp. An important ingredient in Indian cookery for flavouring.

Walnut

THE handsome and durable walnut tree, *Juglans regia*, comes from Asia Minor, ranging from Greece to the Himalayas. It spread from Italy and France to England about 500 years ago. It had been well known in Roman times, but little information has come down to us from classical sources. In England, however, it soon became a common and respected tree; 'walnut avenues' were a feature of many gardens, and it was conventional to have a walnut tree in the stableyard. Valued both for its wood and for its fruit, the walnut has been particularly subject to depredation; the Peninsular Wars saw most of the English trees cut down to provide gun stocks, and a large part of the rest fell to the early nineteenth-century fashion for walnut furniture. The Civil War took most of America's trees. A natural enemy, frost, is always a danger; the winter of 1709 is noted as having killed most of the walnut trees in Central Europe. There has, surprisingly, seemed little interest in maintaining the numbers of this advantageous economic tree. In fruit alone, a good tree should yield 150 lb (75 kg) nuts a year.

The autumnal nut itself – the crinkled kernel of the nut inside the fruit – is a valuable source of protein, although it is also 60 per cent fat. Uniquely flavoured, it has a tendency to form acid, and should therefore be sparingly used. Green walnuts, richer in vitamin C than the riper nuts, can be pickled or made into jam or ketchup. The ripe nuts, familiar in cakes and biscuits, add particularly to sauces and stuffings, and seem to go well with sea food; the French cook a trout *farcie aux noix* which has crushed walnuts with chopped shallots, parsley, breadcrumbs, butter and seasoning as its stuffing. Old wine and walnuts have been familiar after dinner fare since the eighteenth century, and the French liqueur *brou*, known as *nocino* in Italy, is made from the shucks of walnuts. The American walnut, *J. nigra*, is the black walnut, not often available in Europe. It is of greater protein value and can take its place in any recipe for the European walnut, but has a hard, thick shell often requiring special nut-crackers. America also produces the 'white walnut' (the Butternut, *Juglans cinerea*), also hard-shelled.

IRANIAN WALNUT AND CHIVES OMELETTE

Beat 6 eggs in a bowl and mix in 4 oz (100 g) chives, chopped very finely. Stir in 1 oz (25 g) currants and 3 oz (75 g) fresh breadcrumbs, and add a pinch of turmeric and a touch of saffron, with salt and pepper to taste. Mix in 3 oz (75 g) chopped walnuts. Butter and heat a large omelette pan, tip in the egg mixture, and fry; quarter the omelette, turn the segments over and cook lightly on the other side.

And here is a simple Russian version – *yaichnitsa s oryekhami* – of the dish:

WALNUT OMELETTE

Take 4 lightly beaten eggs and whisk in 2 tablespoons cream. Fry 2 oz (50 g) chopped walnuts for a few minutes in butter in a frying pan, then add the eggs, and cook as an omelette.

DEVILLED CRAB AND WALNUTS

Have 2 ½-pint (300-ml) cups cooked crabmeat, flaked, and toss it with 3 oz (75 g) walnuts that have been toasted and chopped and 2 chopped hard-boiled eggs. Stir 2 teaspoons lemon juice into ¼ pint (150 ml) mayonnaise, then stir in a little Worcestershire sauce, a pinch of cayenne pepper, ¼ teaspoon dry mustard and the same amount of salt. Mix all this with the crabmeat, and spoon it into a shallow oven dish. Spread 3 oz (75 g) buttered breadcrumbs over it, and bake in a hot oven (425°F/220°C/mark 7) for 20 minutes. When golden brown, remove, and serve hot with chopped parsley sprinkled on each serving.

WALNUT PIE

Butter a pie dish, put in a layer of mashed potatoes, and season with pepper and salt. Next, chop 4 oz (100 g) shelled walnuts into fine pieces, and cover the potatoes with them. Make ½ pint (300 ml) sauce from brown stock and a little mushroom ketchup and pour it in. Cover with another layer of mashed potato, and bake.

COMPOTE OF WALNUTS

With a knife, split the shells of 'wet' walnuts in half, take out the kernels, peel them without breaking them, then place them in cold water to which the juice of half a lemon has been added. When required for serving, drain them and arrange in a glass dish. Pour over the nuts a reduced apricot syrup (see recipe for peach Melba, p. 203, but use an apricot, instead of a raspberry purée), and cover with stiff whipped cream, flavoured with orange curaçao.

HONEY WALNUT TARTLETS

Sift 6 oz (175 g) flour with a pinch of salt and rub in 4 oz (100 g) butter. Blend in with the beaten yolk of 1 egg. Let the paste cool, then roll it out and line tartlet tins. Mix equal quantities – 5 tablespoons of each would be quite adequate – of honey, Demerara sugar and finely chopped walnuts – and fill the tartlets. Bake for 5 minutes at 400°F/200°C/mark 6, and then for 5 minutes at 350°F/180°C/mark 4 and remove to cool when brown and crisp.

PIZZA FIGLIATA DI NAPOLI

Make shortcrust pastry (see p. 11), and roll it out on a board evenly and very thin. Brush it over lightly with honey, then strew over it chopped walnuts, together with a sprinkling of finely chopped hazelnuts, chopped candied peel and mixed spice. Then carefully roll up the pastry into a long sausage shape and twist it into a spiral. Bake in a moderately hot oven (400°F/200°C/mark 6) for half an hour, until it is a golden brown colour: the spiral can be cut into 2 or 3 lengths for convenient fitting on to an oven pan.

NAMES: Gk. *carya persica*; Latin *nux regia*; Fr. *noix*; It. *noce*; Sp. *noesses*; Ger. *Walnuss*.

STORAGE: A few months if the shell is in good condition, and in dry containers; walnuts can be buried in salt for further preservation.

INDEX